Street Mathematics and School Mathematics is a comparison of mathematics used in school and out of school, describing the two forms of activity as different cultural practices that are based upon the same mathematical principles.

Many philosophers and psychologists have recognized that reasoning about numbers and space is part of people's everyday experience as well as part of the formal discipline of mathematics. However, discussions of everyday mathematical reasoning have been speculative, because until the work described here little systematic research had been carried out comparing mathematical knowledge developed in and out of school. *Street Mathematics and School Mathematics* illustrates the advantages and disadvantages of the two practices as they are now observed, pointing out the trade-off between preservation of meaning and potential for generalization in mathematical knowledge. The empirical findings are analyzed within a broad theoretical framework in the concluding chapter, which discusses the educational implications of these findings and presents a case for realistic mathematics education – a form of teaching that builds formal mathematical knowledge on the foundations of street mathematics.

Street mathematics and school mathematics

Learning in doing: Social, cognitive, and computational perspectives

GENERAL EDITORS: ROY PEA, *Institute for the Learning Sciences*
JOHN SEELY BROWN, *Xerox Palo Alto Research Center*

Street mathematics and school mathematics

TEREZINHA NUNES
University of London
ANALUCIA DIAS SCHLIEMANN
Federal University of Pernambuco
DAVID WILLIAM CARRAHER
Federal University of Pernambuco

CAMBRIDGE
UNIVERSITY PRESS

Published by the Press Syndicate of the University of Cambridge
The Pitt Building, Trumpington Street, Cambridge CB2 1RP
40 West 20th Street, New York, NY 10011-4211, USA
10 Stamford Road, Oakleigh, Victoria 3166, Australia

© Cambridge University Press 1993

First published 1993

Printed in the United States of America

Library of Congress Cataloging-in-Publication Data
Nunes, Terezinha.
Street mathematics and school mathematics / Terezinha Nunes,
Analucia Dias Schliemann, David William Carraher.
p.   cm. – (Learning in doing)
ISBN 0-521-38116-9(hc). – ISBN 0-521-38813-9 (pb)
1. Mathematics – Social aspects.   I. Schliemann, Analucia Dias.
II. Carraher, David William.   III. Title.   IV. Series.
QA10.7.N86   1993
370'.15'651–dc20   92–23183
                                                              CIP

A catalog record for this book is available from the British Library.

ISBN 0-521-38116-9 hardback
ISBN 0-521-38813-9 paperback

Contents

Preface

The ideas discussed in this book resulted from a research program carried out during approximately 10 years at the Universidade Federal de Pernambuco. During this period, our contact with colleagues from different universities has been extremely helpful. We express our gratitude to them all. We want to register our special recognition of those who continually stimulated and supported us in these 10 years: Peter Bryant, University of Oxford; Jean Lave, University of California; Lauren Resnick, University of Pittsburgh; and Sylvia Scribner, City University of New York. We have lost our cherished colleague Sylvia Scribner since the completion of this manuscript and we mourn her.

We also thank the institutions that supported our research and program of exchange of scholars over these years: CNPq, CAPES/PADCT/SPEC, the British Council, and the Fulbright Commission.

Street Mathematics and School Mathematics is here published for the first time in English. It was preceded by a collection of original research papers in Portuguese under the title *Na vida dez, na escola zero.* (São Paulo: Editora Cortez) and by a Spanish translation of the same collection (México: Siglo Veintiuno Editores).

Series foreword

This series for Cambridge University Press is becoming widely known as an international forum for studies of situated learning and cognition.

Innovative contributions from anthropology; cognitive, developmental, and cultural psychology; computer science; education; and social theory are providing theory and research, which seeks new ways of understanding the social, historical, and contextual nature of the learning, thinking, and practice emerging from human activity. The empirical settings of these research inquiries range from the classroom to the workplace, to the high technology office, to learning in the streets and in other communities of practice.

The situated nature of learning and remembering through activity is a central fact. It may appear obvious that human minds develop in social situations and that they come to appropriate the tools that culture provides to support and extend their sphere of activity and communicative competencies. But cognitive theories of knowledge representation and learning alone have not provided sufficient insight into these relationships.

This series is born of the conviction that new and exciting interdisciplinary syntheses are underway, as scholars and practitioners from diverse fields seek to develop theory and empirical investigations adequate to characterizing the complex relations of social and mental life, and to understanding successful learning wherever it occurs. The series invites contributions that advance our understanding of these seminal issues.

1 What is street mathematics?

I Introduction

Many psychological phenomena are difficult to define. In the identification of an object, where does perception end and cognition start? Where does trial and error end and intelligent adaptation start? When and how do we decide that a child's reading problems are no longer "normal" and must be seen as a "learning difficulty"? The difficulty of working in fields with fuzzy boundaries must not discourage investigation. It is possible to study phenomena with working definitions that are to some extent implicit and that become progressively more explicit with further work. This is the case in street mathematics. What we are calling street mathematics used to be known in the literature as informal mathematics. It was a phenomenon with fuzzy boundaries. Most authors seemed to be using implicit working definitions that were basically obtained by exclusion: What looked like mathematics but was not formal mathematics was informal mathematics.

Not only are the boundaries between formal and informal mathematics fuzzy, but even distinguishing mathematics from other activities sometimes seems embarrassingly difficult. For example, at the 5th International Conference on Mathematics Education, D'Ambrosio (1984) showed several pictures of perfectly symmetrical constructions obtained by Brazilian Amerindians and asked why we did not call them studies in geometry instead of house or boat building. Similarly, at a conference on the psychology of mathematics education, van der Blij (1985) showed a series of slides and asked the mathematics educators present whether they were photographs of works of art or the constructions of a geometer. Many examples were

1

"misclassified" by the audience, and van der Blij asked where art ends and geometry starts.

Ginsburg (1982), in an empirical study of the relationship between informal and formal mathematics, proposed an explicit set of definitions to be used in distinguishing different systems of mathematical knowledge. He classified knowledge of elementary mathematics into three systems. System 1 involved techniques for solving quantitative problems developed by children before entering school and without the support of a numeration system. System 1 was defined as *informal* because it developed outside the school setting, and *natural* because it did not involve information or techniques transmitted by culture. System 2 involved mainly the use of counting to solve various arithmetic problems; it was also defined as *informal* because it developed outside school, but *cultural* because it depended on the social transmission of cultural information. System 3 included written symbols, algorithms, and explicitly stated principles generally taught in school; it was thus termed *formal* and *cultural*. Ginsburg's definition, although helpful, still leaves us with a difficulty. The definition tells us nothing about the nature of informal mathematics itself. He defines "informal" by exclusion–informal systems of mathematical knowledge are those forms of mathematical knowledge not taught in school. We must figure out where the mathematics was learned in order to know whether it is informal or formal.

As studies on informal mathematics start building a corpus of information on how it is used, by whom, and where, a better understanding of what informal mathematics is becomes possible. Several contributions to the literature are important here. These contributions can be grouped into two classes of studies: (*a*) work that aims at describing informal mathematics used in Western cultures, and (*b*) work that aims at describing non-Western indigenous forms of mathematics existing in cultures where no systematic transmission in school prevails.

A substantial portion of the work on *informal mathematics in Western cultures* focuses on young children and elementary arithmetic. Several important contributions to our knowledge of elementary arithmetic in preschool years were made by Carpenter and Moser (1982), Fuson (1982), Ginsburg (1977, 1982), Groen and Resnick (1977), Resnick (1982), and Steffe, Thompson, and Richards (1982). All of these studies clearly demonstrated that when children learn a numeration system and understand it well they can then invent ways of using it to solve arithmetic problems through counting and decomposition. For example, Groen and Resnick (1977) showed that children can easily be taught to solve addition problems by representing all of the elements in the addends and then counting them all. They do not, however, follow this procedure strictly if given a large number of practice trials, but discover that they can count on from the cardinal of one of the addends. This procedure of counting on can be discovered by children without specific instruction and exemplifies Ginsburg's System 2 type of mathematical knowledge.

A second group of studies on informal mathematics in Western cultures focuses on mathematics used outside school by adults, not by children. Among these studies, Scribner's and her co-workers' (see Scribner, 1984a,b,c, 1986; Fahrmeier, 1984) analyses of problem solving in work settings and Lave's (1988) investigations of arithmetic in everyday life have made important contributions. Both lines of investigation have demonstrated that it is one thing to learn formal mathematics in school and quite another to solve mathematics problems intertwined in everyday activities. Whether it is inventory taking at work or shopping or calculating calories in cooking, school mathematics does not play a very important role. Informal mathematics has its own forms that are adaptations to the goals and conditions of the activities under way.

Work on *non-Western mathematics* showed that several groups of people who learn numeracy without schooling use their indigenous counting systems to solve arithmetic problems through counting, decomposition, and regrouping (Gay & Cole, 1967; Ginsburg, 1982; Reed & Lave, 1981; Saxe, 1982) in ways that are similar to what one observes in children from Western cultures. For example, Gay and Cole (1967) report in their pioneer work in this field that the Kpelle people of Liberia used stones as support in solving arithmetic problems and could solve addition and subtraction problems using numbers up to 30 or 40 with accuracy. Beyond that, their method became tedious, and people tended to guess at a larger number rather than work out an exact solution. Similarly, Reed and Lave (1981) reported among unschooled tailors in Monrovia, Liberia, the recourse to buttons and other manipulanda as a support in calculation. Like the Kpelle, tailors who relied on manipulanda had difficulty in calculating with larger numbers because their methods relied fundamentally on counting. Saxe (1982) also reported the use of body parts other than fingers among Oksapmin schoolchildren and unschooled adults engaged in commerce to solve addition and subtraction problems. In this case, subjects learned outside school an indigenous counting system based on the names of body parts and adapted their original nonbase system to counting with a base related to the grouping used in the currency they were exposed to.

Informal mathematics among non-Western people has not been restricted to arithmetic problem solving. Other activities of a mathematical nature have also been investigated. Cole, Gay, Glick, and Sharp (1971) covered quite a wide range of activities like measurement and estimation of length, volume, and time; geometry; logical reasoning; and tachistoscopic estimation of number. Saxe and Moylan (1982) investigated transitive inferences based on measurements taken with non-Western measuring systems in which standards are not fixed. Bishop (1983) investigated spatial representation and the concept of area in geometrical forms and land measurement.

Despite the richness of this data base, there is still much that remains unclear about informal and formal mathematics. In some studies, mathematical activities have been observed as they are embedded in everyday life (Scribner, 1984c, 1986, and Lave, 1988, are outstanding examples). In these cases, mathematics is a means at the service of some other goal. Calculating the best buy in a supermarket is a means toward a choice of products; calculating calories is a means toward losing weight; and so on. Mathematics is embedded in some other activity that gives meaning to the situation as a whole. In these studies, the definition of "informal" precedes observation. Mathematical reasoning may take several forms, and any form observed in everyday activity will be an example of informal mathematics. Several other studies have presented subjects with problem situations created exclusively for looking at problem solving, in a laboratory or school-like situation. In this case, the aim of the subject's activity is to solve the problem, to demonstrate competence to the interviewer. Mathematics in these studies is not a means but an end in itself. Whether subjects relied on formal or informal mathematical processes is determined a posteriori. If the methods used in problem solving are not school methods, they are informal methods.

There is clearly a need for a more explicit definition of informal and formal mathematics. The difficulty in attaining these definitions resides in the fact that a definition would in this case amount to a theory. Characterizing informal mathematics in a particular way is the same as putting forth a theory about informal mathematics, its similarities and differences vis-à-vis school mathematics. But we still lack systematic comparisons of informal and formal mathematics. Much of the work on informal mathematics has not investigated formal mathematics. This book aims at filling this gap in the literature. We start with the working definition that mathematics practiced outside school is informal. We look for conditions that allow us to observe street mathematics and school mathematics. We interview people who are likely to know one type of mathematics better than the other. We try to create conditions that lead the same people into using one or the other form of mathematics. In the end, we attempt to establish connections among the three types of mathematics: the one constructed by children outside school, the one embedded in everyday cultural practices, and the one that school aims to teach in the classroom.

To do all this, we carried out several studies in and out of school to allow for a systematic analysis of similarities and differences between school mathematics and street mathematics. The studies relate to arithmetic and simple mathematical concepts. They cover a wide age range and several types of people–children and adults, students and workers, urban and village people. This variety was sought in order to obtain greater generality in our descriptions of street mathematics. We did not want simply to find out how one type of everyday practice involving mathematics takes place, but how several types of activities involving different problems may have something in common. If there are similarities in the processes of mathematical reasoning across everyday practices of vendors, foremen on construction sites, and fishermen, carpenters, and farmers, we can think of a more general description of street mathematics. Would a general description show that street mathematics is, after all, the same as school mathematics, or would there be a clear contrast? We have chosen the labels "street mathematics" and "school mathematics" because they do not commit us from the start to a particular view about what these two types of mathematics are like; they define the types of mathematics simply in terms of scenario.

II   The significance of street mathematics

Street mathematics is not a curiosity without further consequence. It is a phenomenon of interest to sociologists, anthropologists, educators, and psychologists. Sociologists will be interested, for example, in analyzing the social conditions under which street mathematics appears and what relationship it bears to variables commonly used to describe a society. Anthropologists will be interested in street mathematics as cultural practices that have an organization surpassing the level of the individual (i.e., they are not idiosyncratic) and that are in some way transmitted within the culture. It is of interest to anthropologists whether there is a general phenomenon that can be called street mathematics (like kinship structures or classification systems) or whether street mathematics simply comprises sets of particular routines and scripts learned by people to cope with specific situations. Educators will be interested in such questions as whether the children at a given grade level are likely to know particular mathematical concepts from their experiences outside school, whether the new knowledge they gain in school can increase the power of their knowledge outside school, and whether classroom teaching of a novice and of a street expert should be different. Finally, psychologists will be interested in the organization of knowledge in street mathematics, its forms of representation, its power to generate solutions to problems, and its acquisition. There is a sense in which none of these disciplines can study street mathematics without considering the others. An educator, for example, can hardly investigate whether children at a certain grade level may know about a particular concept without the support of the other three disciplines. Nevertheless, there is a sense in which each of these disciplines can maintain its identity even though all are looking at the same phenomenon, because each raises different types of questions.

This book is mostly about the psychological aspects of street mathematics, but we cannot ignore questions that relate to the other disciplines. For this reason, we start by considering aspects of the other disciplines that are relevant both for researchers who may want to replicate or expand on these studies and for others who may wish to apply ideas stemming from the findings. In this chapter, we look briefly at the sociological conditions that we believe create the opportunity for the development of street mathematics in Brazil. This analysis represents a sort of background for the studies described in the chapters that follow. Anthropological and educational questions will be looked at, albeit briefly, in the context of the different studies.

1. Social factors in the emergence of street mathematics

There are people in many countries who in their everyday lives need mathematical concepts and techniques that they have not learned in school. Who these people are and what they do varies across cultures as a consequence of the differences in school systems and in the structure of society and the job market. We assume that in societies in which schooling is more pervasive and a stronger force in the determination of the job market, official forms of knowledge will take precedence over unofficial forms. More people will be exposed only to official forms of mathematics even if they are more difficult to learn. We also assume that in other societies, in which children are under different types of pressure, children's involvement with street and school mathematics may be different. However, we expect to develop a general approach to street mathematics, one that can deal with similarities and differences between street and school mathematics, even in cultures where street mathematics may be less widespread.

We describe below some social factors that we believe favor the development of street mathematics in Brazil. They relate both to the deficiencies of the local school system in the context of the class stratification of Brazilian society and to the structure of the job market.

a. The school system and class differences

The Brazilian school system is deficient both quantitatively and qualitatively. According to Costa (1984), 26% of Brazilians above the age of 15 were illiterate in 1980. This high level of illiteracy is explained as a result of both quantitative and qualitative deficits. Quantitative deficits are obvious and did not change markedly between 1974 and 1980. In 1974, 25% of the children of compulsory school age (7 to 14 years) were not in school (Knight & Moran, 1981); in 1980, 27% of such children were out of school (Costa, 1984). This percentage was equivalent to four million children of school age being out of school in 1980. The percentage varies across regions in the country, reaching up to 33% in the northeast (Knight & Moran, 1981), where most of our studies were conducted.

Not only are there not enough places for all children of school age, but also the quality of the system – certainly in interaction with other factors – produces markedly poor results. The system is neither able to make the children progress at the appropriate pace, nor is it able to retain them in school. As early as the first year of school, 7% of the children drop out; a further 26% drop out between the first and second years (Costa, 1984). The average rate of failure for first and second grades is 25% (Costa, 1984); that means that these children do not progress at the normal pace of one school grade per year, but get left back. According to Knight and Moran (1981), only 24% of the children entering first grade in 1964 made it to fourth grade by their fourth year of school.

These qualitative and quantitative deficiencies of the Brazilian school system create a further peculiarity in the educational system. Although the average age of first graders entering day school is under 8 years, the average age of children in the subsequent grades increases rather quickly, with the standard deviation also increasing. That means that children at the same grade level in state-supported schools may differ considerably in age and general life experiences but will be exposed to the same school curriculum. In designing our studies, we had to make a choice between controlling for age or for grade level. *If we chose to control for age,* we could either (*a*) sample children from different grade levels and ignore the impact of school instruction on the development of mathematical concepts or (*b*) sample children of the same age from the same grade level, working with a selected group of children in the system who had somehow progressed at the pace of a grade level per year. *We chose instead to use grade level rather than age in the definition of our samples.* By doing so, we obtained at least some reference to the degree of our subjects' exposure to school mathematics.

The difficulty of progressing through the school system at the expected pace has resulted in the emergence of evening programs, to which youngsters and adults may turn if at a later point in life they decide to improve their level of schooling. Evening classes enroll youngsters above age 14 and adults in programs ranging from literacy to secondary school. These youngsters and adults follow basically the same curriculum at school. The same methods often are used, although teaching may be expected to proceed at a different pace (for example, two grade levels may be covered in one school year).

This bleak overall picture of the school system differs dramatically from that in private schools, which in Recife represent 30% of school enrollment in primary education, according to data from the local Board of Education. It is not possible to find official statistics that separate private and state-school failure rates by grade. However, in a specific study of children's knowledge of arithmetic in which children from six private and state schools were interviewed, we observed failure rates in second grade of 2% in private schools and 31% in state-supported schools (Carraher & Schliemann, 1983). Discussing the disparity in the quality of education provided in the different schools, Levin (1984, p. 30) bitterly remarks: "Poor children have no choice but to attend state schools, while the rich send their children to private schools. . . . Since the rich do not need the services of state schools, they can accept their low level of performance."

b. Schooling and the job market

According to Pastore (1982, p. 80),

career beginning in Brazil carries numerous social problems stemming from a greater problem, which is that of a still unequal society. Most Brazilian families are unable to educate their children adequately and to postpone their entry into the labor market. The early entry of the children of the lower classes is in itself a strong indicator of social inequality which historically has permeated the Brazilian social system.

As a consequence of the deficiencies in the school system and the pressures on families to send their children into the job market at an early age, 63% of heads of family in Brazil never finished the first stage of primary school, which comprises four years of schooling (Pastore, 1982).

The impact of schooling on the job market is not negligible. Through several regression analyses, Pastore (1979) evaluated the impact of several social factors in determining the occupational status of Brazilians in 1973, taking into account such variables as age, occupational status of the father, initial status of the individual, migration, and education. For the general job market, education always accounted for the largest portion of the variance in occupational status and remained a significant predictor even after all the other variables had been partialed out.

Despite the significance of this finding, it provides only a general picture. The impact of education on the job market is not homogeneous across levels of education, as Velloso (1975) has pointed out. Analyzing the job market on the basis of census data from 1970, Velloso proposed its stratification into three sectors: secondary, primary routinized, and primary independent. Jobs in the *secondary sector* are characterized by their requiring little on-the-job training and offering low wages, poor work conditions, no job stability, and no possibilities of promotion. Education has an impact neither on productivity nor on wages in the secondary sector. Jobs in the *primary routinized sector* offer slightly better wages. Most of the required abilities are acquired through in-service training. Education has little effect on productivity and wages but serves a selective function along with other factors, such as sex, race, and personal characteristics (e.g., submissiveness, stability, and reliability). Finally, jobs in the *primary independent sector* include formal education requirements, usually at a technical or university level. Education is a determining factor in employment.

Thus, the impact of education on the job varies markedly with the level of education under consideration. Similar conclusions were reached through a different analysis by Velloso (1984) in a later study, in which he looked at the payoff of different levels of educational investment. This payoff, which was already markedly unequal (Levy, 1969), became even more diversified in the 1970s. The payoff of primary and secondary education fell by 40% and 22%, respectively, and the payoff of upper-school and university education rose by 28% and 23%, respectively.

In short, employment conditions in Brazil are such that low-status occupations require no schooling. In the urban environment, illiterates and adults with low levels of schooling are engaged in the informal sector of the economy and in low-security jobs in industry. Employment training in special apprenticeship programs (such as those set up in the 1930s for training in commercial occupations [SENAC] and industrial occupations [SENAI]) is rare and does not have a real impact on the job market. This picture of the job market involves an official denial of the requirement of abilities for carrying out jobs in the secondary and primary routinized sector: These jobs are seen as not involving school abilities or specially developed abilities because there are no formal requirements as such. Despite this social characterization, everyday activities in these sectors require of people skills that are not only not recognized but also skills for which people have no formal training. For example, being a foreman in the construction industry – a job in the primary routinized sector – involves unrecognized skills, like counting bricks in large piles (often with not all the bricks visible), calculating the volume of sand delivered at the construction site, calculating the area of walls built by bricklayers paid by the piece, locating the precise places for the foundations on the site and for walls within the buildings, fixing the height of windows before floors have actually been put in, estimating amounts of work to be done by different types of workers so that they are appropriately shifted around as construction proceeds and no one is left without work, and so on. Several kinds of skills go into this list, including mathematical skills.

Everyday mathematics in the occupations of the secondary and primary routinized sector may be just simple arithmetic or involve mathematical concepts like proportion or directed numbers. In either case, there are often differences between the cultural practices learned in school and those used in everyday activities. The mathematical skills used in everyday activities go unrecognized. They are so embedded in other activities that subjects deny having any skills. We often came across comments like this one by a woman who fished for oysters: "I don't know these things; I didn't go to school. I know about the oysters because we fish; the price of oysters we have to know. If they're selling 5 oysters at 750, then they're selling each one at 150." Similar denials of knowledge have been observed in other cultures when people solve problems in what they consider "their own way" (see Cockcroft, 1986; Lave, 1988). However, these denials should not discourage us from looking further into people's mathematical knowledge. This is what we tried to do in the studies described in the following chapters.

III Plan of the book

The studies in the following chapters analyze the differences and similarities between everyday and school mathematics from the psychological point of view. Chapters 2 and 3 look at working-class children's knowledge of street and school arithmetic practices. Chapter 4 describes the loss and preservation of meaning in calculation when different mathematical practices are used in problem solving. It contrasts carpenters with their school-instructed apprentices, and also farmers with students from the same area, identifying differences in their strategies and calculation routines. Chapters 5 and 6 discuss schemas related to the solution of proportions problems developed by two groups: construction foremen and fishermen. Finally, chapter 7 attempts to sketch some general conclusions from the empirical work considered in the book, putting forth some theoretical interpretations and exploring the educational lessons from this work.

2 Arithmetic in the streets and in schools

I Introduction

The comparison of informal and formal procedures in arithmetic – that is, the way people manipulate numbers in solving addition, subtraction, multiplication, and division – is a natural starting point for these investigations for two main reasons. First, arithmetic is a very simple aspect of mathematics (so simple that, according to D'Ambrosio [1986], it was for a long time not considered mathematics). Thus, it is a likely area for the development of street mathematics. Second, many of the studies on street mathematics have focused on arithmetic, and some important contributions have already been made to the field (Ginsburg, 1977, 1982; Reed & Lave, 1981; Saxe, 1982). These studies indicate that there are often multiple systems of arithmetic in the same culture, one related to the school culture and one that flourishes outside school.

Saxe (1982), for example, identified two types of arithmetic practice among the Oksapmin in Papua New Guinea. One of these is based on the indigenous numeration system that uses body-part names as number labels. Although this numeration system was used initially only in counting, with the introduction of a money economy the system became a resource in calculating too. Both unschooled

We are grateful to Elisabete M. de Miranda and Maria Eneida do R. Maciel for their help with data collection and to Shirley Brice Heath, who, visiting our program, opened the way for the questions we asked in this study. We also acknowledge the permission received from the British Psychological Society to include in this chapter Tables 2.1 and 2.2 and other substantial portions of the paper "Mathematics in the streets and in schools," *British Journal of Developmental Psychology*, 3 (1985), 21–29. The research conducted received support from the Conselho Nacional de Desenvolvimento Científico e Tecnológico, Brasília, and from the British Council. The authors wish to thank Peter Bryant for his helpful comments on the present report.

adults, particularly those involved in trading, and schoolchildren were observed to use the body-part system in arithmetic problem solving. Saxe's observations about the use of this street arithmetic system confirms our notion that street mathematics develops mostly when there is a discrepancy between people's needs in problem solving and the amount of mathematics they have learned in school. Unschooled adults clearly could not draw on the school system, and accordingly were observed using only body-part arithmetic. Schoolchildren, however, received formal instruction on how to solve arithmetic problems using Western algorithms from teachers who did not know the Oksapmin language and numeration system. One could expect that children being instructed in a foreign language and in foreign practices would, at least at the beginning of their school career, resort to their indigenous practices in solving arithmetic problems. As schooling becomes a more important source of knowledge, the indigenous system will tend to drop from the picture. This is just what Saxe (1982, p. 172) observed:

> Many children used the conventional body part system to help them solve the arithmetic problems during the test; however, the frequency with which children used their bodies to solve the different tasks differed over grade level. While the majority of children in Grade 2 used their bodies, by Grade 6, only 10% used their bodies.

Similar observations were obtained by Ginsburg, who studied arithmetic problem solving among two groups in the Ivory Coast. One group, the Dioula, was involved in commercial activities; the other, the Baoule, comprised agriculturalists. Ginsburg expected these groups to have different needs, and thus different levels of practice, for arithmetic problem solving. He observed that the unschooled children in the merchant group displayed more efficient arithmetic procedures than their agriculturalist counterparts. With schooling, both groups eventually tended to adopt written arithmetic procedures.

A third line of studies confirms the notion that street arithmetic develops when people become involved in tasks requiring problem-solving skills that they have not learned in school. Reed and Lave (1981) showed that Liberian tailors who had not been to school solved arithmetic problems in different ways from tailors who had attended school for a significant amount of time. Unschooled tailors tended to use procedures that Reed and Lave termed "manipulation of quantities" strategies. These strategies basically involved representing values with pebbles, buttons, or lines on paper and counting them with the support of the indigenous numeration system. In contrast, schooled tailors tended to use written arithmetic and Western-type algorithms – a type of strategy that Reed and Lave called "manipulation of symbols."

These studies certainly suggest that there are informal ways of doing arithmetic calculations that have little to do with the procedures taught in school. They all documented differences *across groups* as a function of their level of schooling. However, it is quite possible that the same differences between street and school arithmetic could exist *within individuals*. In other words it might be the case that the same person could solve problems sometimes in formal and at other times in informal ways. This seems particularly likely with children, who often have to do mathematical calculations outside school that may be beyond the level of their knowledge of school algorithms. It seems quite possible that these children might have difficulty with routines learned at school and yet at the same time be able to solve in other more effective ways the problems for which these routines were devised. One way to test this idea is to look at children who have to make frequent and quite complex calculations outside school. The children who sell things in street markets in Brazil form one such group and are the focus of this chapter.

We start by describing the informal economy in which working-class Brazilian children are engaged, and its significance for their survival. In the next section, we describe a study that investigated their knowledge of street and school mathematics. In the final section, we raise some hypotheses about the nature of street mathematics, which will be pursued in the next chapter.

II The cultural context

The study was conducted in 1982 in Recife, a city on the northeastern coast of Brazil of which the population was then approximately 1.5 million. Like several other large Brazilian cities, Recife receives a very large number of migrant workers from the rural areas who must adapt to a new way of living in a metropolitan region. In an anthropological study of migrant workers in São Paulo, Berlinck (1977) identified four pressing needs in this adaptation process: finding a home, acquiring work papers, getting a job, and providing for immediate survival (in rural areas the family often obtains food through its own work). During the initial adaptation phase, survival depends mostly upon resources brought by the migrants or received through begging. A large proportion of migrants later become unspecialized manual workers, either maintaining regular jobs or working in what is known as the informal sector of the economy (Cavalcanti, 1978). The informal sector can be characterized as an unofficial part of the economy that consists of relatively unskilled jobs not regulated by government organs and hence producing income not susceptible to taxation but not affording job security or such workers' rights as health insurance. The income generated thereby is thus intermittent and variable. The dimensions of a business enterprise in the informal sector are determined by the family's capacity for work. Low educational and professional qualification levels are characteristic of the rather sizable population that depends upon the informal sector. In Recife, approximately 30% of the workforce is engaged in the informal sector as its main activity and 18% as a secondary activity (ibid.). The importance of such sources of income for families in Brazil's lower socioeconomic strata can be easily understood if we note that the income of an unspecialized laborer's family is increased by 56% through his wife's and children's activities in the informal sector in São Paulo (Berlinck, 1977). In Fortaleza it represents fully 60% of the lower-class family's income (Cavalcanti & Duarte, 1980a).

Several types of occupations – domestic work, street vending, shoe repairing, and other types of small repairs carried out without a fixed commercial address – are grouped as part of the informal sector of the economy. The occupation considered in the present study – that of street vendor – represents the principal occupation of 10% of the economically active population of Salvador (Cavalcanti & Duarte, 1980b) and Fortaleza (Cavalcanti & Duarte, 1980a). Although no specific data regarding street vendors were obtained from Recife, data from Salvador and Fortaleza serve as close approximations because these cities are also state capitals in the same geographical region.

It is fairly common in Brazil for sons and daughters of street vendors to help out in their parents' businesses. From about the age of 8 or 9, the children will often enact transactions for the parents when they are busy with another customer or away on some errand.

Preadolescents and teenagers may even develop their own "business," selling such snack foods as roasted peanuts, popcorn, coconut milk, or corn on the cob. In Fortaleza and Salvador, where data are available, 2.2% and 1.4%, respectively, of the population actively engaged in the informal sector as street vendors were aged 14 or under, and 8.2% and 7.5%, respectively, were aged 15–19 (Cavalcanti & Duarte, 1980a,b).

In their work, these children and adolescents have to solve a large number of mathematical problems, usually without recourse to paper and pencil. Problems may involve multiplication (one coconut costs x; four coconuts, $4x$), addition (4 coconuts and 12 lemons cost $x + y$), and subtraction (Cr\$ 500 $- x$, i.e., 500 cruzeiros minus the purchase price will give the change due). Division is much less frequently used but appears in some contexts in which the price is set with respect to a unit of measure (such as 1 kg) and the customer wants a fraction of that unit–for example, when the particular item chosen weighs 750 g. The use of tables listing prices by number of items (one egg – 12 cruzeiros; two eggs – 24; etc.) is observed occasionally in natural settings but was not observed among the children who took part in the study. Pencil and paper were not used by these children, although they are occasionally used by adult vendors when adding long lists of items.

III The empirical study

1. Subjects

The children in this study were four boys and one girl aged 9–15 with a mean age of 11.2 and ranging in level of schooling from first to eighth grade. One had only one year of schooling; two had three years of schooling; one, four years; and one, eight years. All were from very poor backgrounds. Four of the subjects were attending school at the time and one had been out of school for two years. Four of the subjects had received formal instruction in mathematical operations and word problems. The subject who attended first grade and dropped out of school was unlikely to have learned multiplication and division in school, for instruction in the algorithms for these operations usually starts in second or third grade in public schools in Recife.

2. Procedure

The children were approached by the interviewers on street corners or at markets where they worked alone or with their families. Interviewers chose subjects who seemed to be in the desired age range – schoolchildren or young adolescents – and obtained information about their age and level of schooling along with information on the prices of their merchandise. Test items in this situation were presented in the course of a normal sales transaction in which the researcher posed as a customer. Purchases were sometimes carried out. In other cases the "customer" asked the vendor to perform calculations on possible purchases. At the end of the informal test, the children were asked to take part in a formal test, which was given on a separate occasion no more than a week later, and by the same interviewer. Subjects answered a total of 99 questions on the formal test and 63 questions on the informal test. Since the items in the formal test were based upon questions in the informal test, the order of testing was fixed for all subjects.

a. The informal test

The informal test was carried out in Portuguese in the subjects' natural working situation, that is, at street corners or an open market. Testers posed successive questions about potential or actual purchases and obtained verbal responses. Responses were either tape-recorded or written down by an observer, along with comments. After obtaining an answer for the item, testers questioned the subject about his or her method for solving the problem.

The method can be described as a hybrid between the Piagetian clinical method and participant observation. Interviewers were not merely interviewers; they were also customers – questioning customers who wanted vendors to tell them how they performed their computations.

Here is an example taken from the informal test with M., a coconut vendor aged 12, third grader, where the interviewer is referred to as "customer":

CUSTOMER: How much is one coconut?
M.: Thirty-five.

CUSTOMER: I'd like ten. How much is that?

M.: [Pause] Three will be one hundred and five; with three more, that will be two hundred and ten. [Pause] I need four more. That is . . . [Pause] three hundred and fifteen . . . I think it is three hundred and fifty.

This problem can be mathematically represented in several ways: 35×10 is a good representation of the question posed by the interviewer. The subject's answer is better represented by $105 + 105 + 105 + 35$, which implies that 35×10 was solved by the subject as $(3 \times 35) + (3 \times 35) + (3 \times 35) + 35$. The subject can be said to have solved the following subitems:

(a) 35×10;
(b) 35×3 (which may already have been known);
(c) $105 + 105$;
(d) $210 + 105$;
(e) $315 + 35$;
(f) $3 + 3 + 3 + 1$.

When one represents in formal mathematical fashion the problems solved by the subject, one is in fact attempting to represent the subject's mathematical competence. M. proved to be competent at finding out how much 35×10 is, even though he used a routine not taught in third grade, since in Brazil third graders learn to multiply any number by 10 simply by placing a zero to the right of that number. Thus, we considered that the subject solved the test item (35×10) and a whole series of subitems (b to f) successfully in this process. In the process of scoring, however, only one test item (35×10) was considered as having been presented and therefore correctly solved.

b. The formal test

After subjects were interviewed in the natural situation, they were asked to participate in the formal part of the study, and a second interview was scheduled at the same place or at the subject's house.

The items for the formal test were prepared for each subject on the basis of problems solved by him or her during the informal test. Each problem solved in the informal test was mathematically represented according to the subject's problem-solving routine.

From all the mathematical problems successfully solved by each subject (regardless of whether they constituted a test item or not), a sample was chosen for inclusion in the subject's formal test. This sample was presented in the formal test either as a mathematical operation dictated to the subject (e.g., $105 + 105$) or as a word problem (e.g., Mary bought x bananas; each banana cost y; how much did she pay altogether?). In either case, each subject solved problems employing the same numbers involved in his or her own informal test. Thus, quantities used varied from one subject to another.

Two variations were introduced in the formal test according to methodological suggestions contained in Reed and Lave (1981). First, some of the items presented in the formal test were the inverse of problems solved in the informal test (e.g., $500 - 385$ might be presented as $385 + 115$ in the formal test). Second, some of the items in the informal test used a decimal value that differed from the one used in the formal test (e.g., 40 cruzeiros may have appeared as 40 centavos, or 35 may have been presented as 3,500 in the formal test). (The principal Brazilian unit of currency is the cruzeiro; each cruzeiro is worth 100 centavos.)

In order to make the formal test situation more similar to the school setting, subjects were given paper and pencil and were encouraged to use them. When problems were nonetheless solved without recourse to writing, subjects were asked to write down their answers. Only one subject refused to do so, claiming that he did not know how to write. It will be recalled, however, that the school-type situation was not represented solely by the introduction of pencil and paper but also by the very use of computation exercises and word problems typical of school practice.

In the formal test, the children were given a total of 38 mathematical operations and 61 word problems. Word problems were rather simple, and each involved only one mathematical operation.

3. Results and discussion

Analysis of the results from the informal test required an initial definition of what would be considered a test item in that situation. Whereas in the formal test items were defined prior to testing, in the informal test problems were generated in the natural setting and items were identified a posteriori. In order to avoid a biased increase in the number of items solved in the informal test, the defi-

Table 2.1. *Results according to testing conditions*

| | Informal test | | Formal test | | | |
| | | | Arithmetic operations | | Word problems | |
Subject	Score	No. of items	Score	No. of items	Score	No. of items
M	10	18	2.5	8	10	11
P	8.9	19	3.7	8	6.9	16
Pi	10	12	5.0	6	10	11
MD	10	7	1.0	10	3.3	12
S	10	7	8.3	6	7.3	11
Totals		63		38		61

Note: Scores are adjusted to a 10-point scale.

nition of an item was based upon questions posed by the customer/ tester. This probably constitutes a conservative estimate of the number of problems solved, because subjects often solved a number of intermediary steps in the course of searching for the solution to the question they had been asked. Thus, the same defining criterion was applied in both testing situations in the identification of items even though items were defined prior to testing in one case and after testing in the other. In both testing situations, the subject's oral response was the one taken into account even though in the formal test written responses were also available.

Problems presented in the streets were much more easily solved than ones presented in school-like fashion. We adjusted all scores to a 10-point scale for purposes of comparability. Table 2.1 shows the adjusted scores obtained by each child in each situation. The overall percentage of correct responses in the informal test was 98.2 (in 63 problems solved by the 5 children). In the formal-test word problems (which provide some descriptive context for the subject), the rate of correct responses was 73.7%, which should be contrasted with a 36.8% rate of correct responses for the arithmetic operations.

A Friedman two-way analysis of variance of score ranks compared the scores of each subject in the three types of testing conditions. The scores differ significantly across conditions ($x^2r = 6.4$, $p = .039$).

Mann-Whitney U's were also calculated comparing the three types of testing situations. Subjects performed significantly better on the informal test than on the computation exercises ($U = 0, p < .05$). The difference between the informal test and the word problems was not significant ($U = 6, p > .05$).

It could be argued that errors observed in the formal test were related to the transformations that had been performed upon the informal test problems in order to construct the formal test. An evaluation of this hypothesis was obtained by separating items that had been changed either by inverting the operation or changing the decimal point from those items that remained identical to their informal test equivalents. The percentage of correct responses in these two groups of items did not differ significantly; the rate of correct responses in *transformed* items was slightly higher than that obtained for items identical to informal test items. Thus, the transformations performed upon informal test items in designing formal test items cannot explain the discrepancy of performance in these situations.

A second possible interpretation of these results is that the children interviewed in this study were concrete in their thinking and that concrete situations would thus help them in the discovery of a solution. In the natural situation, they solved problems about the sale of lemons, coconuts, and so on with the items in question physically present. However, the presence of a concrete instance can be understood as a facilitating factor if the instance somehow allows the problem solver to abstract from the concrete example to a more general situation. There is nothing in the nature of coconuts that makes it easier to discover that three coconuts (at Cr$ 35.00 each) cost Cr$ 105.00. The presence of the groceries does not simplify the arithmetic of the problem. Moreover, computation in the natural situation of the informal test was in all cases carried out mentally, without recourse to external aids for partial results or intermediary steps. One can hardly argue that mental computation is an ability characteristic of concrete thinkers.

The results seem to be in conflict with the implicit pedagogical assumption of mathematical educators according to which children ought first to learn mathematical operations and only later apply them to verbal and real-life problems. Real-life and word problems may provide the "daily human sense" (Donaldson, 1978) that will guide children to find a correct solution without requiring an extra step –

namely, the translation into arithmetic sentences. This interpretation is consistent with data obtained by others in the area of logic – for example, Johnson-Laird, Legrenzi, and Legrenzi (1972), Lunzer, Harrison, and Davey (1972), and Wason and Shapiro (1971).

How is it possible that children capable of solving a computational problem in the natural situation will fail to solve the same problem when it is taken out of its context? In the present case, a qualitative analysis of the protocols suggested that the problem-solving routines used may have been different in the two situations. In the natural situations, children tended to reason by using what can be termed a "convenient group," whereas in the formal test school-taught routines were more frequently, although not exclusively, observed. Five examples are given in Table 2.2 that demonstrate the children's ability to deal with quantities and their lack of expertise in manipulating symbols. The examples were chosen to represent clear explanations of the procedures used in both settings. In each of the five examples, the performance described in the informal test contrasts strongly with the same child's performance in the formal test when solving the same item.

In the informal test, children rely upon mental calculations that are closely linked to the quantities being dealt with. The preferred strategy for multiplication problems seems to consist in chaining additions. In the first example, as the addition became more difficult the subject decomposed a quantity into tens and units – to add 35 to 105, M. first added 30 and later included 5 in the result.

In the formal test, where paper and pencil were used in all the examples in Table 2.2, the children tried, without success, to follow school-prescribed routines. Mistakes often occurred as a result of confusing addition routines with multiplication routines, as is clearly the case in examples (1) and (5). Moreover, in all the cases, there is no evidence, once the numbers are written down, that the children try to relate the obtained results to the problem at hand in order to assess the adequacy of their answers.

Summarizing briefly, the combination of the clinical method of questioning with participant observation used in this project seemed particularly helpful when exploring mathematical thinking and thinking in daily life. The results support the thesis proposed by Luria (1976) and by Donaldson (1978) that thinking sustained by daily human sense can be at a higher level than thinking out of

Table 2.2

---

(1) First example (M., age 12)

Informal test

CUSTOMER: I'm going to take four coconuts. How much is that?
CHILD: There will be one hundred five, plus thirty, that's one thirty-five . . . one coconut is thirty-five . . . that is . . . one forty!

Formal test

Child solves the item 35 × 4, explaining out loud: "Four times five is twenty, carry the two; two plus three is five, times four is twenty." Answer written: 200.

(2) Second example (MD, age 9)

Informal test

CUSTOMER: OK, I'll take three coconuts [at a price of Cr 40.00 each]. How much is that?
CHILD: [Without gestures, calculates out loud] Forty, eighty, one twenty.

Formal test

Child solves the item 40 × 30 and obtains 70. She then explains the procedure: "Lower the zero; four and three is seven."

(3) Third example (MD, age 9)

Informal test

CUSTOMER: I'll take twelve lemons [one lemon is Cr$ 5.00].
CHILD; Ten, twenty, thirty, forty, fifty, sixty [while separating out two lemons at a time].

Formal test

Child has just solved the item 40 × 3. In solving 12 × 5, she proceeds by lowering first the 2, then the 5 and the 1, obtaining 152. She explains this procedure to the (surprised) examiner when she is finished.

(4) Fourth example (S., age 11)

Informal test

CUSTOMER: What would I have to pay for six kilos [of watermelon at Cr$ 50.00 per kg]?
CHILD: [Without any appreciable pause] Three hundred.
CUSTOMER: Let me see. How did you get that so fast?
CHILD: Counting one by one. Two kilos, one hundred. Two hundred. Three hundred.

Formal test

Test item: A fisherman caught 50 fish. The second one caught five times the amount of fish the first fisherman had caught. How many fish did the lucky fisherman catch?

Table 2.2 *(cont.)*

CHILD: [Writes down 50 × 6 and 360 as the result; then answers] 36.
Examiner repeats the problem and child does the computation again, writing down 860 as result. His oral response is 86.
EXAMINER: How did you calculate that?
CHILD: I did it like this. Six times six is thirty-six. Then I put it there.
EXAMINER: Where did you put it? [Child had not written down the number to be carried.]
CHILD: [Points to the digit 5 in 50] That makes eighty-six [apparently adding 3 and 5 and placing this sum in the result].
EXAMINER: How many did the first fisherman catch?
CHILD: Fifty.

(5) Fifth example

Informal test
CUSTOMER: I'll take two coconuts [at Cr$ 40.00 each. Pays with a Cr$ 500.00 bill]. What do I get back?
CHILD: [Before reaching for customer's change] Eighty, ninety, one hundred. Four twenty.

Formal test
Test item: 420 + 80.
The child writes 420 plus 80 and claims that 130 is the result. [The procedure used was not explained, but it seems that the child applied a step in a multiplication routine to an addition problem by successively adding 8 to 2 and then to 4, carrying the 1; that is, 8 + 2 = 10, carry the 1, 1 + 4 + 8 = 13. The zeros in 420 and 80 were not written down. Reaction times were obtained from tape recordings; the whole process took 53 sec.]
EXAMINER: How did you do this one, four twenty plus eighty?
CHILD: Plus?
EXAMINER: Plus eighty.
CHILD: One hundred, two hundred.
EXAMINER: [After a 5-sec pause, interrupts the child's response, treating it as final] Hmm, OK.
CHILD: Wait a minute. That was wrong. Five hundred. [The child had apparently added 80 and 20, obtaining 100, and then started adding the hundreds. The experimenter interpreted 200 as the final answer after a brief pause, but the child completed the computation and gave the correct answer when solving the addition problem by a manipulation-with-quantities approach.]

context in the same subject. They also raise doubts about the pedagogical practice of teaching children how to solve mathematical operations simply with numbers before using the operations in the context of problems.

Our results are also in agreement with data reported by Lave, Murtaugh, and de la Rocha (1984), who showed that problem solving in the supermarket was significantly superior to problem solving with paper and pencil. It appears that daily problem solving may be accomplished by routines different from those taught in schools. In the present study, daily problem solving tended to be accomplished by strategies involving the mental manipulation of quantities. Whereas in the school-type situation the manipulation of symbols carried the burden of computation, thereby making the operations "in a very real sense divorced from reality" (see Reed & Lave, 1981, p. 442). In many cases, attempts to follow school-prescribed routines seemed in fact to interfere with problem solving.

IV   Further questions

The results of this study can be looked at in different ways. There is one sense in which they are not surprising: We searched for *street mathematics*, and we found what we looking for. The examples we observed were rather similar to those already reported in the literature. But there is more to these results than just finding one more instance of street mathematics: the marked within-individual differences documented here.

Within-individual differences represent a difficulty for most psychological theories. Trait theories, which expect achievement in mathematics to be linked with giftedness, have no way of predicting large within-individual differences when the same arithmetic operations are being tested for. Nor can cognitive development theories of the structuralist type handle this type of data. If two tasks involve the same cognitive operations, there should be no appreciable gap in the solutions. Despite the coining of the term "horizontal décalage" in Piagetian theory, there is nothing in the theory that would allow one to expect children to perform radically differently in the arithmetic tasks we have been discussing – unless we can say that the two types of question rely on different logical structures. Finally, behavoristic approaches would leave the matter up to the notion of transfer, under the supposition that what is learned in one situation may not transfer to a new situation. Despite the apparent reasonableness of this explanation, one should not forget that the children in this study had received formal teaching in the methods they were failing in. Transfer should be a matter of using what was learned in school in the street situation rather than the other way around.

In the search for some theoretical explanation that could help us understand this phenomenon, we did find one theory that can accommodate these findings and, what is more important, can lead to further hypotheses about the differences and similarities between school and street mathematics: the theoretical framework proposed by Vygotsky and Luria. Luria (1979, p. 6; authors' translation) summarized their basic ideas in the following passage:

The psychological life of animals has its origin in their activity and it is a form of representation of reality; it is carried out by the brain but it can only be explained by the laws of the activity of representation. Similarly, the higher forms of conscious activity, directed attention, memory, and logical reasoning, cannot be considered as a natural product of the evolution of the brain but result from the specific form of social life which is characteristically human.

Scribner and Cole (1981, pp. 8–9) further explored this idea by considering that

mental processes always involve signs, just as action on the environment always involves physical instruments (if only the human hand). Changing tools alters the structure of work activity: Tilling the soil by hoe and by tractor require different patterns of behavior. Similarly, Vygotsky claimed, changing symbol systems restructure mental activity. According to Vygotsky, basic psychological processes (abstraction, generalization, inference) are universal and common to all humankind; but their functional organization (higher psychological processes, in Vygotsky's terminology) will vary, depending on the nature of the symbol systems available in different historical epochs and societies and the activities in which these symbol systems are used.

This theoretical formulation allows for the prediction of both *across-groups* and *within-individual* differences, which arise from the same circumstances: the support of mental activity by different systems of representation. As we worked with the children in the streets and as we look at the literature on informal arithmetic, one fact strikes us: Street arithmetic is oral and school arithmetic is written. Could the fact that problem solving in street and in school mathematics is supported by different types of symbolic representation explain the within-individual differences we observed? Can we find enough differences in the "functional organization" of oral and written arithmetic to account for the observed discrepancy in performance? Or, alternatively, is it possible that these two types of system are based on different logical principles? These are the questions we propose to analyze in the next chapter.

3   Written and oral arithmetic

I   Introduction

The study presented in the preceding chapter leaves several possible explanations for the differences in performance observed in the informal and formal tests. In particular, two types of theory could explain these results – one stressing the social-interaction aspects of the situations and a second stressing the social-cognitive aspects. A social-interactionist position would maintain that the situation in which mathematical calculations are performed determines the role participants have with respect to each other, and thus influences people's success. Arithmetic problems that arise in the marketplace are an indispensable part of a commercial transaction between vendor and customer. It may be that the social relationship established between vendor and customer is such that children feel more confident of their own ability and perform better. They trust themselves as vendors but not as students, and this lack of confidence makes their performance deteriorate when they have to act as students. A social-cognitive interpretation would go along lines suggested by Luria and Vygotsky and look at the impact of the symbolic systems used in the two different situations rather than at the interaction between the people. Perhaps because the types of symbolic system used in street mathematics and school mathematics are different – one being oral and the other written – children structure their activities in different ways; this would explain the difference in their performance.

We wish to acknowledge the permission received from the National Council of Teachers of Mathematics, USA, to reprint Tables 3.1–3.7 and other substantial portions of the paper "Written and oral mathematics," which appeared in the *Journal for Research in Mathematics Education, 8*, (1987), 83–97.

On the basis of the study described in chapter 2 alone, it is not possible to decide whether a social-interactionist or a social-cognitive hypothesis is a better explanation for those findings, because there were important differences between the two conditions under which children were interviewed. The relationship between the experimenter and the subject in the two settings was different, and the arithmetic system used by the children across situations was also different. From the street vendors' point of view, the interviewer was a customer – perhaps a bit more talkative than a typical customer, but a customer nonetheless. There was thus no reason for the youngsters to become anxious or inhibited during the "testing," as might be the case in the school-like situation. Such differences could conceivably explain the superior performance demonstrated by the youngsters when solving problems in the market as compared to the school-like setting. Although we did not observe any overt signs of anxiety when the youngsters were tested in the school-like setting, the confounding in experimental design is still there. Thus the results are open to different interpretations. What we needed in order to sort out these possibilities was a replication of the differences between street and school arithmetic but in the same situation. We needed a study in which the relationship between the experimenter and the child remained constant, but problem solving supported by oral versus written practices of arithmetic was still observed.

This chapter describes a study in which we attempted just that: to elicit the different types of arithmetic under a constant relationship between experimenter and child. By interviewing children in only one setting, we could maintain constant the physical and social setting as well as the relationship between the experimenter and the child. This is of course a risky proposition, because if the physical setting and the social-interaction factors were crucial for the emergence of street mathematics, we should fail to obtain the very phenomenon we wanted to study. On the other hand, if we could simulate the conditions that provoke the use of oral arithmetic, we could come closer to understanding the results of the first study.

From the study of street and school mathematics we had learned that working-class children are exposed to two different practices of arithmetic, oral and written, and that the contexts in which these two occur tend not to overlap. Oral arithmetic is the practice observed in street markets and shops. Written arithmetic could be observed when we presented children with school-like exercise, like word-problem and computation exercises. Why not simulate it all, then, in one interview?

II The empirical study

The first important step to take was to locate schools that catered mostly to working-class children in areas where we knew there were street markets. We identified two suitable schools, one in the middle of a shantytown and one in a mixed neighborhood. Both schools were located in plazas where street markets were held on Fridays and Saturdays. In both schools, the algorithms for addition and subtraction were taught in second grade, and those for multiplication and division were taught in third grade. For this reason, we chose third graders as our target population. We told the teachers we were interested in how children solve arithmetic problems and asked them to allow us to interview children from their classes. We also asked to be introduced simply as people interested in how children solved problems. The children's perspective was likely to have been that of one being tested or interviewed by an "outside teacher."

In this general context, we asked the children to solve arithmetic problems in three conditions: (*a*) in a simulated store condition, in which the child played the role of the storekeeper and the experimenter played the role of customer, (*b*) embedded in word problems, and (*c*) as computation exercises. In the simulated store condition, concrete objects – such as cars, dolls, and marbles – were spread about the table, but there was no money available. The word problems were of the simplest types (see, for example, Vergnaud, 1982, for an analysis of the complexity of addition and subtraction word problems, and Brown & Burton, 1978, for multiplication and division). Since our interest was in the arithmetic as such, not in the children's understanding of relationships in the problem, we wanted to minimize the difficulty of problems. The computation exercises consisted of simple instructions to carry out an arithmetic operation, like "What is 200 minus 35?"

1. Experimental design

Each child was asked to solve 10 problems in each of the three conditions described above. Testing was carried out preferably

Table 3.1. *Number combinations used across testing conditions*

Set	Combinations				
A	60 + 240	115 + 15	195 + 57	40 × 3	12 × 50
	200 − 35	210 − 105	143 − 68	100/4	75/5
B	115 + 15	420 + 80	210 − 105	15 × 50	4 × 25
	500 − 80	195 + 57	252 − 57	120/3	100/4
C	200 − 35	210 − 105	80 + 420	3 × 40	50 × 12
	106 + 106	185 + 68	243 − 75	75/5	100/4

during a single session but occasionally had to be extended to two sessions because of the school's routine (e.g., end of the school day or beginning of snack time). Paper and pencil were available on the desk at all times, but the children were free to use whatever procedure they wanted when solving problems.

The arithmetic computations were matched across subjects for the different conditions so that no differences in performance across conditions could be explained by formal differences between the problems. Table 3.1 shows the three sets of number combinations used. Table 3.2 lists the word problems and the simulated store problems in which the number combinations from Table 3.1 were used.

The order of conditions followed a Latin square design for each block of three children; in this way, neither practice nor fatigue nor a particular bias caused by the children's initial approach to the problems could explain any differences found among the experimental conditions. Five replications per cell were planed, but one extra child was inadvertently tested in one of the cells, and we decided not to discard the results of that interview.

By studying how the same children's performance varied across the different experimental conditions, we expected to find out whether the type of social interaction was a crucial determinant in eliciting the street and school mathematics practices we had observed in the previous study.[1] By controlling the order of conditions and the numbers used in problems, we could assure that any differences obtained among the experimental conditions were not due to different levels of difficulty in the computations carried out or to fatigue, practice, or

[1]The social relationships with the child was constant across testing situations; it was a teacher–child relationship. It could not explain within-subject variation across testing conditions.

Table 3.2. *Problems presented in the simulated shop and as word problems*

No.	Statement

Word problems*[a]*

1   Mark went to see a movie. He spent *x* cruzeiros for the bus and *y* for the movie ticket. How much did he spend altogether?

2   I bought a mango for *x* cruzeiros. I paid with a *y*-cruzeiro bill. How much change did I get?

3   In a school there are *x* classrooms. In each classroom there are *y* children. How many children are there in this school?

4   Maria gave *x* cruzeiros to *y* boys who washed her car. They divided the money so that each one had the same amount as the others. How much money did each boy get?

5   John had *x* marbles. He played with Paul and won *y* marbles. How many marbles does he have now?

6   I had *x* soccer cards in my collection. I lost *y*. How many do I have now?

7   Robert had *x* marbles. He played with a friend and lost *y*. How many does he have now?

8   Peter bought *x* eggs. Each egg cost *y* cruzeiros. How much money did he spend?

9   Mr. Roger gave *x* marbles to *y* boys to share among themselves. Each one should get the same amount as the others. How many marbles did each boy get?

10   Jack bought a ball for *x* cruzeiros and a car for *y*. How much money did he spend altogether?

Simulated shop*[b]*

1   One ring costs *x* cruzeiros. One piece of candy costs *y* cruzeiros. I want one of each. How much do I have to pay you?

2   The pen costs *x* cruzeiros. I am paying you with a *y*-cruzeiro bill. How much is my change?

3   Each pencil costs *x* cruzeiros. I want *y* pencils. How much do I have to pay?

4   You're selling *x* cars for *y* cruzeiros, but I only want one. How much does one car cost?

5   I want to buy this pen that costs *x* cruzeiros and this ball that costs *y*. How much do I have to pay?

6   Let's say this doll costs *x* and this pencil costs *y*. I'm buying both. How much do I have to pay?

7   You had *x* pencils. You sold *y*. How many do you still have in your store?

8   I have *x* cruzeiros in my pocket. I want to buy this bag of marbles with it. You're selling the bag for *y*. How much money will I have left?

9   *x* of these toy guns cost *y* cruzeiros. How much will you sell one of them for?

10   You're selling each of these pencils with erasers for *y*. I want *x* of them. How much will I have to pay?

*[a]*Numbers were inserted in word problems and in the problems presented in the simulated shop according to the design described in the text.
*[b]*The interviewer always referred to objects on the table.

increased familiarity with the experimenter. Thus, if we successfully elicited both practices of arithmetic and replicated the differences in rate of success observed previously, we would find support for the social-cognitivist position put forth by Luria and Vygotsky, who maintain that the symbolic systems used in mental activities influence their functional organization and consequently their efficiency.

2. Subjects

The subjects were 16 third graders ranging in age from 8 to 13 years, with a mean age of 11.5 years, randomly selected from two state-supported schools in Recife, Brazil. As mentioned in the preceding chapter, the wide age range observed in this study is typical of state schools. The younger children in the study had succeeded in learning how to read in one or two years and had started at age 7. The older children either started school late or repeated first or second grade one or more times.

All children had received school instruction in algorithms for adding, subtracting, multiplying, and dividing, including long division, on which they had been instructed 2 months before the beginning of the study.

3. Procedure

The children were interviewed individually in their schools. All questions were presented orally. The children were asked to explain how they found their answers when their procedure for solving the problem was not clear. They were encouraged to speak out loud if they spontaneously subvocalized. Children who attempted to solve a particular problem with pencil and paper yet *from their own viewpoint* failed to find a satisfactory answer were encouraged to try to solve the problem in their heads, and vice versa. However, for purposes of statistical analysis, only the child's first attempt was scored.

A description of each child's problem-solving behavior in all three conditions was obtained by means of tape recordings made during the sessions, detailed notes taken by an observer, and written material produced by the children. For each item, the procedure used by the child in solving the problems was classified as either written or oral, and the response was scored as either right or wrong.

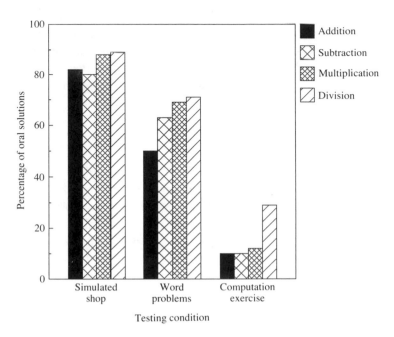

3.1   Percentage of oral procedures in each condition, by type of operation

4. Results and discussion

The results were analyzed in several steps. First, we wanted to know whether we had successfully elicited oral practices in the simulated shop and written practices in the school-type exercises. Next we looked at the effects of condition on performance in terms of the accuracy of responses. Finally, we looked at the structure of the solutions obtained. These steps are described in the following sections.

a. Choice of procedure

The effect of problem condition on choice of procedure can be ascertained by looking at Figure 3.1. It is clear that the children had a marked preference for oral procedures in the simulated shop and for written procedures when solving the computation exercises, with the word problems remaining intermediary between the other two conditions.

The statistical significance of these differences was analyzed by ranking across conditions the relative preference of each child for oral

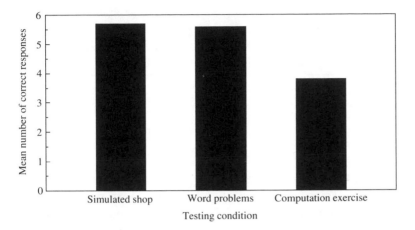

3.2 Mean number of correct responses (out of 10), by testing condition

procedures and performing a Friedman two-way (Subjects × Condition of testing) analysis of variance by ranks. This analysis yielded a significant effect of testing condition on choice of procedures, $\chi^2(2, N = 16) = 16.9, p < .001$. Thus, we were able to simulate the conditions that favor the emergence of street and school arithmetic procedures despite the fact that the social-interaction factors were controlled for. This successful simulation allows us to conclude that physical settings (street and school) and the particular type of relationship (customer–vendor or teacher–pupil) are not crucial factors leading to choice of procedure. It is possible to say that there are culturally accepted practices that take place in and are evoked by these settings but that are not conditioned by them. The same practices can also be evoked in simulated situations – although not all simulations are necessarily successful (see Lave's, 1988, comments on this issue).

b. Testing condition, procedure chosen, and accuracy of responses

Testing condition not only influenced choice of procedure but also had a striking effect on the accuracy of responses (see Figure 3.2). The differences in performance across conditions were evaluated by means of a one-way ANOVA with repeated measures that treated testing condition as a main effect and number of correct responses as the dependent variable (Table 3.3). The significant

Table 3.3. *Summary of the ANOVA with condition as main effect and number of correct responses as dependent variable*

Source	df	Mean square	$F(2, 30)$
Condition	2	18.405	12.78*
Subjects	15	18.854	
Interaction	30	1.440	

* $p < .01$

effect ($p < .01$) was accounted for by the difference between the computation exercises on the one hand and the word problems and simulated store problems on the other, for there was no difference between the mean number of correct responses in the simulated shop and word problems.

The fact that performance did not differ between the simulated shop and the word problems indicates that the availability of concrete objects is not crucial to the better performance displayed under these testing conditions, in contrast to the computation exercises. Whereas objects were present and could in principle be manipulated in the simulated store, this was not the case in word problems. Further, no child manipulated objects as a support to problem solving in addition, subtraction, or division problems. In five of the multiplication problems solved in the store condition (15.6%), four of which were correctly solved, the children did use the concrete items to monitor calculation. (Further reference to monitoring follows in the analysis of oral multiplication procedures.)

According to our hypothesis, the effect of condition on performance was to be explained in terms of the symbolic systems the children had used to support their reasoning. We expected that children would perform differently across conditions, because they had used either oral or written symbolic systems when carrying out their arithmetic. Thus, we looked at the percentage of correct responses according to the procedure used by the children – oral versus written – for each arithmetic operation regardless of testing condition. Figure 3.3 displays the results. In all cases, oral arithmetic was associated with a greater probability of success than written arithmetic, though this association was less marked for addition.

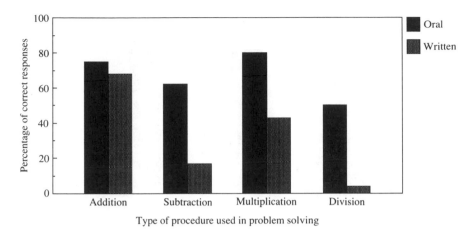

3.3   Percentage of correct responses, by operation and procedure

Given the striking differences between oral and written computation, a further analysis was carried out to test the significance of the difference in correct responses as a function of the type of procedure chosen – oral or written. The type of arithmetic was treated as an independent variable; the dependent variable was the percentage of correct responses observed when each form of arithmetic was used. A Sign Test revealed that the oral procedure was significantly superior to the written procedure at the .002 level.

The effects of testing condition and type of arithmetic are, of course, confounded in the preceding analysis. In order to study the effect of type of arithmetic independently of testing condition, the percentages of correct responses were calculated for all children, with testing condition (simulated store, word problems, and computation exercises) and type of arithmetic (oral versus written) treated as independent variables. Figure 3.4 displays the results of this analysis for all four operations taken together, and Table 3.4 displays these percentages separating out the four operations. There is a clear trend of better performance on oral than on written arithmetic. The significance of this difference was not tested statistically for two reasons: (*a*) The use of either arithmetic practice was not the result of a random assignment to a condition but rather was determined by the child; and (*b*) the number of responses in some cells was quite small

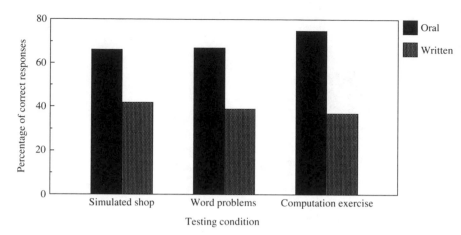

3.4   Percentage of correct responses, by procedure used in problem solving and by testing condition

Table 3.4. *Percentage of correct responses for each operation by type of procedure and testing condition*

	Testing condition		
Procedure	Simulated shop	Word problems	Computations
	Addition		
Oral	67	83	100
Written	75	62	70
	Subtraction		
Oral	57	69	60
Written	22	22	14
	Multiplication		
Oral	89	64	100
Written	50	50	39
	Division		
Oral	50	50	50
Written	0	0	7

(there were only 24 written responses in the simulated shop and 20 oral responses in the computation exercises).

Two alternative explanations still remain for the difference in the observed rates of success between oral and written arithmetic. The first would simply take the difference to reflect diverse amounts of practice. Perhaps working-class children have many opportunities to practice oral arithmetic but few to practice written arithmetic. This hypothesis would hold that the more drill, the greater the efficiency. We cannot in fact count the number of occasions on which these children were drilled in either oral or written arithmetic. However, two considerations suggest that the simple amount of drill is a poor candidate as an explanation here. First, drill is quite systematic at school, and written arithmetic is much more likely to have been the subject of drill than oral arithmetic. Second, the difference in rates of success between the operations also weakens this possibility. Although subtraction must have been the subject of drill much more often than multiplication because it was taught earlier, it remained harder.

A second alternative hypothesis could be based on the notion that we are treating as general differences in success rates across types of arithmetic practice an effect that actually results from individual differences. It is possible that children who were better calculators resorted to oral arithmetic and those who were weaker turned to written procedures. We can test this hypothesis by looking at children's preference for written arithmetic and correlating this preference with the number of problems solved correctly. If the children who resort to written arithmetic do so because they are weaker in calculation abilities, we should find a negative correlation between preference for written arithmetic and number of correct responses.

The number of problems solved by means of written procedures varied between 6 and 28 out of a total of 30 problems solved by each child. The average number of problems solved by written procedures for all conditions was 13.6. The children were ranked according to their preference for written arithmetic as measured by the percentage of total items solved with paper and pencil. Percentages were used because some children had refused to try some division items. This measure of preference for written procedures was correlated by means of the Pearson coefficient with the children's score for correct responses, yielding a positive, nonsignificant result ($r = .06$). This analysis allows us to set aside the explanation in terms of individual differences.

Despite all our attempts to eliminate alternative hypotheses in the explanation of the difference in performance across type or arithmetic practice, two interpretations still remain possible: (*a*) The children

had relatively less difficulty with oral procedures (or the oral procedures are more efficient); and (*b*) the children chose to solve through
oral procedures those problems they had less difficulty with. Since
the children spontaneously decided on the solution procedure, they
may have resorted to school algorithms for the more difficult problems. Finding no way of evaluating these two possibilities through
quantitative analysis, we turn now to a qualitative analysis of the
structure of solutions in oral and written arithmetic practices.

c. Structure of solution in oral and written arithmetic

This final analysis was qualitative, centering on the search
for generality in oral arithmetic practices – despite the fact that they
have hitherto been treated as idiosyncratic (see Cockcroft, 1986;
Hunter, 1977; Plunkett, 1977) – and on the identification of similarities and differences between oral and written arithmetic. This analysis will be described in two steps. In the first section we describe the
procedures that were observed in oral arithmetic. In the second section we compare oral and written arithmetic practices.

Oral arithmetic procedures. Oral arithmetic gives the impression of such
variety of approaches that it seems, at first glance, almost impossible
to classify oral procedures into simple categories. The general impression conveyed is that children are breaking problems into subproblems, but there appears to be no rule about how the subproblems
are chosen. Children sometimes split a number into parts and at
other times work with that same number without splitting it. However,
if we look not at the choice of parts but at the ways in which the parts
are manipulated, we can readily identify two basic types of strategy:
(*a*) decomposition, which is used in addition and subtraction, and
(*b*) repeated grouping, which is used in multiplication and division.

Decomposition: Some examples of decomposition are presented in Table 3.5. Some of the children's remarks quoted here
were spontaneous, and others were prompted by questions posed by
the experimenter. The descriptions include the child's identification,
the experimental condition, and the number combination used in the
problem. The children's oral responses are in italics; our comments
on heuristics are in brackets.

Table 3.5. *Some examples of decomposition*

Example 1

Subject: L. Condition: Word problem. Computation: 200 − 35.

L: *If it were thirty, then the result would be seventy. But it is thirty-five. So it's sixty-five; one hundred sixty-five.* [The 35 was decomposed into 30 and 5, a procedure that allows the child to operate initially with only hundreds and tens; the units were taken into account afterward. The 200 was likewise decomposed into 100 and 100; one 100 was stored and the other was used in the computation procedure.]

Example 2

Subject: E. Condition: Store. Computation: 243 − 75.

E.: *You just give me the two hundred* [he meant 100]. *I'll give you twenty-five back. Plus the forty-three that you have, the hundred and forty-three, that's one hundred and sixty-eight.* [Instead of operating on the 243, the child operated on 100, subtracted 75 and added the result to 143, which had been set aside.]

Example 3

Subject: Ev. Condition: Computation exercise. Computation: 252 − 57.

Ev: *Take fifty-two, that's two hundred, and five to take away, that's one hundred and ninety-five.* [The child decomposed 252 into 200 and 52; 57 was decomposed into 52 + 5; removing both 52s, there remained another 5 to take away from 200.]

The decomposition heuristic, as can be seen, deploys the child's knowledge of the number system. L. decomposed 35 into 30 and 5; it is easier to operate with 100 minus 30 than with 100 minus 35. By decomposing 35 into 30 and 5, she dealt initially with hundreds and tens, leaving units for a later step in the computational process. E. decomposed 243 into 100 and 143, subtracted 75 from 100, and subsequently added the difference to the remaining part of the number.

Decomposition tends to involve rounding numbers. Often this rounding of numbers occurs before computation – e.g., 35 decomposed into 30 (round number) and 5, or 243 decomposed into 100 (round number) and 143. In other cases the decomposition results in a round number after an operation has been performed; when Ev. was solving 252 minus 57, she subtracted 52 from both numbers, a move that led to 200 minus 5 as the next step in computation. Round numbers were actively sought by the children apparently because (*a*) they are more likely to be related to known number facts (100 − 75 is likely to be known) and (*b*) they help avoid the overload in processing that would result from operating simultaneously on hundreds, tens, and units. By decomposing a number, the child can operate successively with the different relative values.

Decomposition was used at least once by each child in both addition and subtraction. Of the 141 additions and subtractions solved orally by all children in the three conditions considered jointly, it was possible to identify 53 clear examples of decomposition (approximately one third). There were 88 responses for which the children did not provide an explanation of how they achieved the result (often merely restating the result as if it were just obvious) and 5 others that clearly were not obtained through decomposition. Thus, when it was possible to identify the procedure used in oral addition and subtraction, the most likely route was decomposition. Of the responses in addition and subtraction obtained through decomposition, 85% were correct.

If we reflect about decomposition and try to identify the logic implicit in the children's reasoning, we can identify the following principles (or theorems in action, to use Vergnaud's, 1982, terminology):

> – A number is composed of parts that can be separated without altering the total value.
> – Addition and subtraction can be carried out on these parts without affecting the final result.

A formal representation of this principle can be provided as:

$$\text{If } a + b = c \text{ and } d + e = f,$$
$$\text{then } (a + d) + (b + e) = c + f.$$

These principles correspond to the property of associativity of addition and subtraction, a property that is also the basis of the written algorithms taught in school (see Resnick, 1986, for a similar analysis of an American child's invented procedures for adding and subtracting). This qualitative analysis indicates that children who use oral arithmetic understand the very principles implicit in the algorithms for column addition and subtraction. Despite this competence, they are baffled by the way written arithmetic is to proceed, especially in the case of subtraction.

Table 3.6. *Some examples of repeated grouping*

Example 1

Subject: JG. Condition: Store. Computation: 15 × 50.
JG: *Fifty, one hundred, one fifty, two hundred, two fifty.* (Pause) *Two fifty. Five hundred, five fifty, six hundred, six fifty, seven hundred, seven fifty.* [The child was monitoring the number of 50s on one hand. When he reached five 50s, he doubled that number, obtaining ten 50s, and then went on counting in single 50s.]

Example 2

Subject: F. Condition: Word problem. Computation: 75/5.
F.: *If you give ten marbles to each [child], that's fifty. There are twenty-five left over. To distribute to five boys, twenty-five, that's hard.* (Experimenter: That's a hard one.) *That's five more for each. Fifteen each.* [The problem was solved by successively subtracting the convenient groups distributed while keeping track of the increasing share that each child received; 10 marbles each were given to 5 children, which accounted for 50 marbles, and the remaining 25 were distributed among the 5 children – 5 to each, totaling 15 for each child.]

Example 3

Subject: Ev. Condition: Computation exercise. Computation: 100/4.
[After attempting unsuccessfully to solve the exercise on paper, Ev. claimed that it was impossible. She first attempted to divided 1 by 4, which was not possible, then to divide 0 by 4, and finally gave up. The examiner asked for a justification.]
Ev.: *See, in my head I can do it. One hundred divided by four is twenty-five. Divide by two, that's fifty. Then divide again by two, that's twenty-five.* [She proceeded here by factoring; two successive divisions by 2 replace the given division by 4.]

Repeated groupings: Let us turn now to some examples of repeated groupings to explore the characteristics of this procedure. The examples presented in Table 3.6 illustrate its flexibility. (In this table, the experimenter's remarks are in parentheses.)

Repeated grouping is a procedure appropriate for multiplying and dividing. It consists of multiplying by means of successive additions or dividing by means of successive subtractions. When the child is multiplying, the values added may be convenient chunks that are easier to operate with than the given multiplicands. Through constant monitoring, the child keeps track of intermediary products and of progress toward a solution. Computation is finished when the proper number of "times" has been added or when the original quantity has been fully distributed in division. Concrete objects or fingers may be used in the monitoring process.

The specific chunks chosen seem to depend on both the numbers involved and the child's knowledge of number facts. Grouping fifties into hundreds brings an obvious advantage to the process of computing; it is very easy to count by hundreds. But the procedure is richer yet: Children can take shortcuts such as (*a*) working out the subtotal of 6 chunks and then doubling the answer to obtain 12 chunks or (*b*) adding 5 more to 10 chunks to obtain 15 or (*c*) making successive doublings – 2, 4, 8, 16 – and then subtracting 1 chunk to obtain 15 chunks. In division, the heuristic involves successive subtractions, as in F.'s approach to 75 divided by 5. Successive divisions, as observed in Ev.'s example, were included here as a case of the repeated-grouping heuristic even though the process of factoring may be more complex: It involves the coordination of successive divisions (or multiplications) instead of successive subtractions (or additions). Further investigation is needed to clarify the issue.

The repeated-groupings procedure was used at least once by 15 of the 16 children in multiplication and by 14 in division. Of the 113 multiplications and divisions solved orally in the three conditions taken jointly, it was possible to identify 61 clear examples of repeated grouping (approximately one half). Twelve responses were attempts at decomposing, one of them mixed with repeated grouping. For the remaining 40 responses, no classification was possible, owing to lack of explanation of the procedure. Of the problems solved by repeated grouping, 59% were correct.

If we try to make explicit the theorems in action that seem to be implicit in children's solutions of multiplication problems through repeated grouping, we can recognize the following:

> – A number can be decomposed into parts without changing in value.
> – These parts can be multiplied by the same number and the products then added, resulting in a value that is the same as what would be obtained if the two numbers were directly multiplied.

A formal representation of these principles would be:

If $a + b = c$, then
$(a \times d) + (b \times d) = c \times d.$

The formal representation of the repeated-grouping procedure makes clear why it differs from decomposition: The logic implicitly involved here is the property of distributivity, which pertains to multiplication and division, rather than that of associativity, which is implicit in decomposition procedures. Distributivity is also the property we use implicitly when we carry out column multiplication. This qualitative analysis points to the same conclusions we came to after looking at decomposition: Children who use oral arithmetic understand the principles that structure algorithms for column multiplication and division. Despite this competence, they do not easily master the school-taught algorithms for multiplication and division.

Oral and written symbolic forms in arithmetic practices. Analysis of the principles implicit in oral and written arithmetic shows that the radically different performances displayed by the same children in different conditions cannot be explained by differences in the logico-mathematical requirements of the tasks. Thus, we shall now turn to the hypothesis derived from Vygotsky and Luria's ideas: It is possible that these discrepant performances can be explained in terms of the symbolic systems being used. If oral and written arithmetic differ in terms of how they are functionally organized, we expect not only such discrepancies in performance but also differences in the structuring of children's activities at the symbolic level. These differences should lead us to find further effects on performance of symbolic systems.

If we look again at the protocols in Tables 3.5 and 3.6, we can identify some important differences between oral and written arithmetic.

First, oral and written procedures differ in the direction of calculation: The written algorithm is performed working from units to tens to hundreds, whereas the oral procedure follows the direction hundreds to tens to units. As Reed and Lave (1981) have pointed out, this means that errors in the lower values become inputs for the higher values in written arithmetic. On these grounds, one should expect written arithmetic to result in larger errors than oral arithmetic.

Second, in the oral mode, the relative value of numbers is preserved: We say "two hundred and twenty-two." In the written mode, the relative value is represented through relative position: we write 2 2 2. This difference in signifiers is maintained during calculation: Oral procedures preserve the relative values; written procedures

Table 3.7. *One child's calculation in written and oral procedures*

Subject: R. Situation: Store. Computation: 200 − 35.

[The child writes 200 − 35 in vertical arrangement. Then he writes the result from units to tens to hundreds, computing out loud, and obtaining 200 in the following way.]

R: *Five, to get to zero, nothing. Three, to get to zero, nothing. Two, take away nothing, two.*

Experimenter: (Is it right?)

R: *No. So you buy something from me, and it costs thirty-five, you pay with a two-hundred-cruzeiro note and I give it back to you?*

E: (Do it again, then.)

[R. writes down 200 − 35 in the same way, writes the result from units to tens to hundreds, computing aloud and obtaining 235 as follows.]

R: *Five, take away nothing, five. Three, take away zero, three. Two, take away nothing, two. Wrong again.*

E: (Why is it wrong again?)

R: *Now you buy something and its costs thirty-five. You give me two hundred and I give you two hundred, and thirty-five on top?*

E: (Do you know what the result is?)

R: *If it were to cost thirty, then I'd give you one seventy.*

E: (But it is thirty-five. Are you giving me a discount?)

R: *One hundred and sixty-five.*

---

set them aside. This is illustrated in the protocol presented in Table 3.7, which describes one child's efforts to solve the same problem in the written mode, without success, and then in the oral mode.

The protocol clearly illustrates in the same subject the difference between the two procedures. When the child attempted to solve the problem in writing, calculation proceeded from smaller to larger, the relative value was set aside, and the wrong answer was repeatedly obtained. In the oral mode, the correct answer was found through steps that proceeded from larger to smaller and preserved the relative value of numbers during the process of calculation. The difference in procedures is all the more striking when the ease with which the child solves the computation in the oral mode is compared with his difficulty in the written mode. The fact that we are looking at the same child, not at different children, strengthens the notion that the differences in performance in oral and written arithmetic are related to the nature of the practices, not to giftedness or level of understanding.

These two characteristics of oral and written practices – direction of calculation and loss of meaning in the written mode – lead us to

Table 3.8. *Percentage of correct responses and errors of different magnitudes as a function of type of procedure*

	Correct	Within 10%	Within 20%	>20% off
Addition				
Oral	74	14	8	4
Written	64	2	4	30
Subtraction				
Oral	69	6	14	11
Written	17	17	4	61

predict that the types of errors that result from the two practices will differ. Responses obtained in the written mode may not be evaluated for sense when calculation is completed and may just remain wrong despite a large margin of error. This prediction can be tested by a simple error analysis, which is summarized in Table 3.8. Responses resulting from oral and written additions and subtractions were divided into four categories: (*a*) correct, (*b*) wrong and within 10% of the correct answer, (*c*) wrong and within 20% of the correct answer; and (*d*) wrong and more than 20% off. We then carried out for each operation a separate evaluation of the association between type of procedure – oral or written – and type of error. For both operations, the results of this analysis indicated a significant association between type of procedure and type of error ($\chi^2 = 17.72$, with $df = 2$, $p < .001$ for addition and $\chi^2 = 16.75$, with $df = 2$, $p < .001$ for subtraction, both calculated with Yates's correction for continuity).

III   Conclusions and further questions

The study described in this chapter shows that the differences across situations observed in the preceding study cannot be explained only by social-interactional factors. Despite the fact that we controlled for the type of interaction between the experimenter and the children by presenting all the problems in one situation, we were able to elicit oral arithmetic practices in a simulated shop and written arithmetic practices in response to computation exercises that are commonly encountered in the classroom. Large discrepancies in performance were still observed across testing conditions, but these effects seemed to be due to the type of arithmetic practice used. When type of arithmetic and testing condition were treated as independent variables, there was variation in performance across practices but not across testing conditions.

Qualitative analyses of the oral and written practices suggest that these practices are based on the same implicit logicomathematical principles, the property of associativity in the case of addition and subtraction and the property of distributivity in the case of multiplication and division. Thus, the discrepancies in performance cannot be explained in terms of diverse logicomathematical principles underlying the different practices.

The most salient differences between the two practices were observed in the way the two symbolic systems used structured the activity of calculation. In contrast to written arithmetic, oral arithmetic proceeds from larger to smaller and works in ways that preserve the relative value of numbers. These two characteristics led to the prediction that errors in oral practice would be smaller than those generated in written arithmetic, a prediction that was supported by further statistical analysis.

The analyses carried out in this study encouraged us to look further at street arithmetic, because further hypotheses can be raised at this point. For example, the problems presented in this study were rather simple in structure; their solutions required a single computation. Not much was required of the children in terms of analysis of problem situations. Still, it was possible to notice that the children kept the meaning of the problem in mind while solving problems in the oral mode and seemed to forget about this meaning quite often when solving problems in the written mode. This was most noticeable in problems with multiplication and division, in which the children repeatedly referred to the items in the problem while calculating in the oral mode (see Table 3.6). If we bear in mind that mathematics involves not just arithmetic calculation but first and foremost the analysis of relationships between variables, we can hypothesize that further differences may be observed in written and oral practices. Is it possible that the relationships between variables also become less visible once pupils turn to written mathematics? This is a question we explore in the next chapter.

4 Situational representation in oral and written mathematics

I Introduction

The comparison of oral and written arithmetic practices presented in the preceding chapter indicates a need to search for ways of analyzing cognitive performance not so far explored in the literature. Piagetian theory, which has inspired research on cognitive development for many decades, has concentrated on the analysis of differences in reasoning that reflect differences in the conceptual invariants. This theory has provided a framework for looking at important conceptual invariants, like conservation, seriation, quantification of class inclusion, proportionality, and so on, but it cannot shed any light on differences in performance in cognitive tasks that are supported by the same invariants. In the last two decades, however, researchers have become aware that perhaps the conceptual invariants are not the only thing that matters for cognitive performance. Perhaps the ways in which concepts are formed are also of significance. In particular, studies of the effects of Western-type schooling on cognitive development have resulted in the idea that school concepts and everyday concepts about the same conceptual field may not be the same.

With respect to street and school mathematics in Brazil, we have already identified some important differences in the type of representation used. Street mathematics is oral and preserves much of the meaning of the situations at hand. Mathematical practice in school is written and leaves out as much of the specifics of situations as possible in striving for generality. In this chapter we want to explore the differences between street and school mathematics further both

The work on carpenters and apprentices was supported by research grants from CNPq and INEP. Ana Karina Lira, Clara Santos, Robério Melo, and Solange Canuto participated in the data collection and analysis.

theoretically and empirically. In the second section of the chapter, different views of the impact of schooling on cognitive development will be discussed briefly. In the third and fourth sections, empirical studies will be presented that contrast performance in mathematical tasks by people who learned most of their mathematics outside school with that of students, whose knowledge of mathematics was definitely influenced by school practices. The last section discusses the findings in a broader perspective and puts forward some conclusions.

II   Different views of schooling and cognitive development

Studies on cognitive development, such as Greenfield (1966), Luria (1976), Rogoff (1981), and Sharp, Cole, and Lave (1979), have shown that schooled people perform better than nonschooled people in a variety of cognitive tasks. However, schooling effects are not always observed. Lave (1988), for example, did not observe schooling effects on everyday mathematics in her project with adults in California, although there was an effect on their mathematical abilities in a school-type test. Schooling may influence general cognitive development even though it may not always give people an advantage in cognitive tasks. We still do not know what predictions to make in particular cases. Further, the mechanisms through which the effect of schooling on cognitive development operates are poorly understood so far.

Different hypotheses have been put forth about the origin of these effects. A first line of explanation is cognitive in nature. Some authors (Bruner, 1966; Vygotsky, 1962) propose that the effect of schooling on cognitive development is related to the means of representation used in school. Bruner, for example, argues that school learning is distanced from reality; it is carried out in the context of general representations of situations. It is this distance of school learning from reality that makes it more abstract and consequently more general than learning in context. Vygotsky also argues that school learning leads to more general and abstract thinking because it is mediated by language rather than stemming from representations of reality. He contrasts concepts that he calls "natural," which are developed from representations of reality in more inductive forms, with concepts he terms "scientific," which are learned from general symbolic definitions later applied to concrete situations. According to Vygotsky, school instruction induces a type of generalizing perception of objects and events in children and creates hierarchical systems of relations between concepts. Scientific concepts formed in this fashion interact with spontaneous concepts to produce a continual interplay between learning and development. Saxe and Posner (1983, p. 297) thus expanded on and clarified Vygotsky's position using today's language of information-processing theories:

> To explain this process from Vygotsky's perspective, it is important to distinguish between two types of learning experiences, those that occur from the "bottom up," giving rise to what Vygotsky has called spontaneous concepts, and those that occur from the "top down," producing what Vygotsky has called scientific concepts.
>
> Bottom-up learning is described as resulting from the child's spontaneous attempt to understand aspects of social and physical reality without the direct aid of practical concepts; that is, the child achieves local solutions to particular problems. . . . In contrast, top-down learning is described as resulting from interactions with adults or more capable peers. In these interactions, problems are posed for the child, and he or she is presented with concepts of general applicability that are valued in the culture. Top-down learning, such as that encountered in school, gives the child the opportunity to form general concepts that may be adapted to different problem types but are not necessarily grounded in immediate experience.

Luria (1976) explored further the influence of schooling (and other cultural activities) on perception and cognition, proposing that learning in context, without the influence of certain cultural experiences, leads people to react directly to what he terms the "graphic perceptual situation," whereas schooling promotes the mediation of abstract systems of representation in perception and cognition.

Scribner and Cole (1973) have added further strength to this line of argument through their description of school learning in contrast to informal learning. They describe school learning as based on verbal explanations and involving little direct demonstration. Informal learning is characteristically learning in context, by observation and imitation, and rarely involves any verbal explanations.

This first line of explanation hypothesizes the existence of two types of learning with different consequences for what is learned and how. School learning, based on symbolic representations distanced from the concrete situations they represent, results in more abstract and generalizable learning. Learning in context, out of school, being based on more direct representations of the situation, results in specific acquisitions that are of less generalizability.

A second approach to understanding the advantages exhibited by schooled over unschooled people is sociocognitive (Doies & Mugny,

1981; Perret-Clermont, 1980; Piaget, 1926, 1976). Schools create social situations that put children's concepts in check. In school, children are often confronted with problems they cannot solve, with situations in which their solutions are not successful, with different views about the same problem situations. These conflicts create opportunities for learning. Piaget, Perret-Clermont, and Doise and Mugny all stress that interactions must be cooperative for children to learn and that learning comes from conflict and reflection, not from imitation of behavioral models. In their constructivist approach, learning processes in school are not distinct from those observed out of school; they are just socially motivated. Piaget stresses that "general coordinations are the same, whether they occur in inter- or intra-individual actions" (1976, p. 226). Similarly, Perret-Clermont (1980) emphasizes that by studying social interactions she is only investigating how interactions influence the facility or rapidity of the growth of operations; she considers the elucidation of the source and the originating mechanisms of operations as well as their structure to be one of Piaget's greatest contributions to psychology.

Thus, according to this second line of explanation, schools are special places created for learning. Learning may be stimulated in schools by social interactions, but whatever learning happens in school is not of a new type.

The preceding chapters on oral and written mathematics are suggestive of yet a third possibility. Learning in school and out of school is connected to specific problem-solving practices – a point made quite strongly by Lave (1988). The differences between the problem-solving practices in each arena have an impact on what type of learning takes place and how knowledge is used. Schooled subjects, although receiving general cognitive benefits from the fact that they are in special places for learning several different subjects, do not always fare better than unschooled subjects. Further, amount of schooling is not always associated with improvement in performance. When more specific levels of analysis are used, it is possible to make finer distinctions between school and out-of-school practices used in problem solving and to arrive at a better understanding of how learning mathematics in school and out-of-school may be different.

We saw in Chapter 3 that the *verbal/nonverbal* distinction does not apply to the practice of mathematics in school versus on the street.

The distinction between school arithmetic and street-market arithmetic is that between *written and oral* practices.

Mathematics problem solving in school is written and relies on procedures that distance it from meaning. This distance from meaning is an advantage in the sense that the forms used in school are more general. Procedures used for whole-number arithmetic, for example, are just as good with decimal numbers.

In contrast, oral practices preserve meaning. If decimal numbers are involved in the situations, new questions are posed for the problem solver. For example, if in a problem involving money one has to add $2.25 and $3.90, the written arithmetic solution is a matter of writing down the numbers and proceeding as if there were no decimal point. As long as the numbers are aligned according to the decimal point, the addition algorithm works just as usual. In oral arithmetic, $2.25 can be read as "two dollars and 25 cents" or "two dollars and a quarter." When "three dollars and 90 cents" are added to this amount, a conversion from cents to dollars is required. The answer "5 dollars and 115 cents," which would be obtained by adding dollars and cents independently, is inappropriate. Because oral arithmetic requires conversions or recoding from cents to dollars, it has explicitly to take into account the changes of units. Taking changes of unit into account keeps the procedures close to meaning and involves more than algorithmic procedures in arithmetic: It involves knowledge of the meaning of the symbols.

Similarly, if measures such as centimeters and meters are used in arithmetic and decimals are involved, oral practices do not proceed as if the situation simply involved rules for calculating with whole numbers plus rules for placing the decimal point. The changes of units of centimeters into meters, for example, must be explicitly recognized. In oral arithmetic, 1.5 m and 1.4 m are not referred to as "one point five" or "one point four" but rather as "one and a half meters" and "one meter and forty centimeters." This oral form of representation keeps subjects close to meaning, whereas written procedures are distanced from meaning – just what we observed for whole numbers in the preceding chapters.

Rules for calculating in written arithmetic can be applied without knowledge of what changes of unit take place after the decimal point. The algorithms comprise explicit rules for how to write the numbers, how to carry out the calculation, and how to place the decimal point, and these rules apply equally well to money, meters, or any other decimal system of measurement. Performing the algorithm requires no understanding of the meaning of the symbols; for this reason, Resnick (1982) calls this type of knowledge syntactic.

In contrast, oral practices demand semantic knowledge; that is, subjects have to know how many cents make a dollar or how many centimeters make a meter in order to calculate. In this sense, oral practices require understanding of the specific quantities involved in the problem, whereas written practices can be of a rule-based nature and can take place in the absence of understanding specific aspects of the situation.

Thus, the differences between oral and written practices are not differences in how abstract the procedures are. Both procedures are of a general and abstract nature. The differences seem to stem from the amount of knowledge required of the subject about the specific situation being handled. There seems to be a trade-off in oral and written arithmetic as far as preservation of meaning and generalizability are concerned. To a greater preservation of meaning in oral arithmetic procedures correspond restrictions either in the range of numbers or in the range of situations that a subject can handle. Similarly, to the loss of meaning in written procedures corresponds a greater range of situations and numbers that can be handled without modification in procedures.

The distance from meaning created in written arithmetic practice can have negative consequences. Students may carry out computations in a proper fashion (demonstrating an ability to handle a great range of numbers) but then fail to interpret results appropriately. Some assessments of children's performance in problem solving have shown that, even when school algorithms are adequately learned, children may end up giving inappropriate answers to arithmetic problems. One example, quoted by Schoenfeld (1985), is found among the 13-year-old students participating in the third National Assessment of Educational Progress (Carpenter, Lindquist, Matthews, & Silver, 1983). When given the problem "An army bus holds 36 soldiers. If 1,128 soldiers are being bussed to their training site, how many buses are needed?" 70% of the 45,000 students who were tested correctly performed the long-division algorithm. When asked to state the answer to the problem, however, 29% answered that the number of buses needed was "31 remainder 12" and 18% gave "31" as the answer. Lave (1977) and Reed & Lave (1981), working with Liberian tailors, found even stronger evidence for the loss of meaning in written arithmetic as a trade-off for the increase in the range of numbers that could be dealt with. Tailors who learned arithmetic in school could easily deal with large numbers by means of school-taught algorithms but made absurd errors that were overlooked at the end of a problem more often than tailors who learned arithmetic problem solving at the shop; the latter, however had difficulty in calculating with large numbers.

The loss of meaning observed at the end of computation in written practices may also be observed at the outset in more complex problems. The distance between the arithmetic procedures used and the representation of the situation in written arithmetic often involves students in the question of finding the correct operation(s). In contrast, in oral arithmetic the arithmetic relations and the situation are part of the same schema, and the question of finding the correct operation(s) must be viewed in very different terms. For example, giving change to a customer may be a matter of counting the money up from the amount spent as bills are handed out or subtracting the amount spent from the amount paid in. In this case, oral arithmetic allows for additive and subtractive solutions to the same problem, whereas written arithmetic works only with the subtractive solution. The closeness of the representation of the situation and the representation of the arithmetic in the schemas developed in street mathematics simultaneously preserves meaning and allows for greater flexibility in the routes used to a solution. Scribner (1984a), in a study contrasting the performance of product assemblers with that of students, demonstrated clearly this greater flexibility of routes to solution when problem representation includes specifics from the situation. Product assembly involves taking an order, which is presented in terms of number of cases of the product, and collecting the amount of the products needed to satisfy the order. An order may be presented as "1 − 6 quarts," which means one case (16 quarts) minus 6 quarts; that is, 10 quarts. Assemblers may fill the order by removing 6 quarts from a case or by using partly full cases; for example, removing 1 quart from a case that has only 11 quarts to begin with. Flexibility in problem solving in this example involves transforming a literal interpretation (16 − 6) into an equivalent nonliteral solution

(11 − 1). Scribner observed that nonliteral solutions were used in order to save physical effort and were substituted for literal solutions whenever these required greater physical effort. A comparison between experienced product assemblers, who had a good picture of the specifics of the situation, and students, who could obviously see the equivalence of nonliteral solutions but did not have such good representations of the specifics of the situation, showed that product assemblers used nonliteral solutions much more often than students when these represented a saving in physical effort. Expert product assemblers used 72% nonliteral solutions, whereas students used only 25% under these conditions. Scribner's study, however, involved rather simple problems in which subjects could correct their solutions and always end up with correctly filled orders. Thus, although she showed that representations developed in practical settings were more flexible, she did not observe errors as a result of representations that were more distanced from the situation.

We now look at two studies that contrast street and school mathematics and concentrate on the analysis of loss versus preservation of meaning in the representation of the problem situation and in the interpretation of the final answer. Both studies involve more complex problem situations than those described up to now. The complexity may be related either to the problem situation itself or to the use of decimal numbers or to both. The first study (Grando, 1988) contrasts the problem-solving efforts of farmers with low levels of schooling with those of students in the same rural areas who were learning school mathematics. The second study (Schliemann, 1984) analyzes problem solving by a group of professional carpenters with low levels of schooling and a group of carpentry apprentices learning mathematics in school. These samples offer an interesting opportunity for comparing street and school mathematics because both contrasting groups are familiar with the contents of the problem situations but learned to solve arithmetic problems predominantly either at work or through formal school instruction.

III   Fieldwork in mathematics: Farmers and students

Grando (1988) interviewed 15 farmers (level of schooling ranging from none to 7 years of school attendance), 20 fifth graders from rural schools, and 20 fifth graders plus 20 seventh graders from urban schools. Rural and urban schools differ in how elementary education is structured: For rural schools, elementary education ends at Grade five, whereas in urban schools elementary education ends at Grade eight. All subjects lived in the same district (Campinas do Sul) in the south of Brazil. Farmers were contacted through friends and were interviewed on their farms. Students were contacted through their teachers and interviewed in school.

The choice of problems for inclusion in the study was based on previous interviews carried out with farmers of the region in which they were asked what arithmetic calculations they had carried out that day, and how and when they had used mathematics at work. They were then asked specific questions about measurements (for example, how they measured land, volume, distances, etc.) and how they carried out the specific calculations they had mentioned in the interview. From these initial interviews, several problems were developed for later presentation to the four groups of subjects.

Subjects were interviewed individually by an interviewer who presented the problem situations orally (not all farmers could read) and explained the situations as often as needed. Interviews were tape-recorded and transcribed in toto for later analysis. Problem content involved situations occurring on farms that were familiar to all four groups of subjects. The list of problems that will be discussed here is presented in Table 4.1. These problems were chosen for discussion here because of their complexity in comparison with the problems we used in the study of children's arithmetic presented earlier. The three problems with decimal numbers involved simple situations, but decimal numbers pose difficulties in and of themselves. The complex multiplicative-relations problem involved simple numbers, but more than one operation was required for solution.

These two types of problem situations will be analyzed differently here. In the simple situations with decimal numbers, difficulties arose mainly when the final result had to be interpreted, although incorrect choices of operations and inversions in the direction of division were also observed among students. These problems will be discussed only in terms of the loss or preservation of meaning in the final response. The complex multiplicative-relations problem involved a situation with more than three variables and will be analyzed in terms of the arithmetic relations represented in the strategy used in problem solving.

Table 4.1

(*a*) *Problems with decimals*
1. A farmer was going to build a gate and had to cut a piece of wire 7 m long into pieces 1.5 m long in order to fit the gate. How many pieces would he get for the gate? (Reasonable range of responses: 1 to 7 pieces.)
2. Some fruit trees were going to be planted in lines 50 m long. Each tree has to stand 2.75 m from the next tree in the same line. How many trees will be planted in a line? (Reasonable range of responses: 1 to 50 trees.)
3. One liter of wine fills 1.5 bottles. How many liters of wine are used to fill 7.5 bottles? (Reasonable range: 1 to 7.5.)

(*b*) *Complex multiplicative-relations problem*
Suppose this is a piece of land [shows a rectangle drawn to scale where the measurements 60 m × 30 m are indicated for length and width]. When you plant tea bushes, the space between the bushes has to be 3 by 4 meters. How many bushes will be planted in this plot?

---

a. Calculating with decimals

Results in the three problems with decimals showed that farmers had a strong preference for oral arithmetic, and students had a strong preference for written arithmetic, demonstrating that farmers used street mathematics most often and that students had a marked preference for school mathematics (see Table 4.2). The farmer's preference for oral arithmetic does not mean that they did not know written arithmetic. Note that Problem 2 was solved through written calculation by most farmers, probably because of the numbers used in the problem.

These differences in arithmetic practice were not without consequence for the type final answer given by the subjects. For each problem, Grando determined intervals within which responses were considered reasonable in view of the meaning of the problem. For example, in the problem involving 7 m of wire to be cut in pieces 1.5 m long, the reasonable range went from 1 (no cuts being made, there would be one piece) to 7 pieces. (If subjects simply ignored the decimal point and cut the 7 m into pieces of 1 m, they would come up with the answer "seven pieces.") She then noted the percentage of responses inside and outside the reasonable range for each group of

Table 4.2 *Percentage of solutions through written or oral in three problems with decimals*

Subjects	Written	Oral	No attempt
	Problem 1		
Fifth grade, rural	75	25	0
Fifth grade, urban	90	10	0
Seventh grade, urban	80	18	2
Farmers	20	80	0
	Problem 2		
Fifth grade, rural	85	15	0
Fifth grade, urban	95	05	0
Seventh grade, urban	100	0	0
Farmers	85	15	0
	Problem 3		
Fifth grade, rural	80	20	0
Fifth grade, urban	80	20	0
Seventh grade, urban	70	25	5
Farmers	13	87	0

Source: Grando, 1988.

subjects (ranges are presented in Table 4.1). The percentages by group of subjects are presented in Figure 4.1.

Both groups of fifth graders gave significantly more responses outside the reasonable range than farmers, according to tests for the significance of independent proportions (for the contrast with rural fifth graders, $z = 3.15$, $p < .001$; for the contrast with urban fifth graders, $z = 2.02$, $p < .05$); the difference between farmers and seventh graders was not significant ($z = 1.58$, $p > .05$). Grando further observed that the extreme responses given by students were often the result of an appropriate operation followed by failure to place the decimal point correctly. For example, in the problem in which 7 m of wire had to be divided into pieces 1.5 m long, students' answers varied from 0.4 to 413 pieces (both of the named answers obtained by dividing 7 by 1.5).

Grando's results confirm the idea that street mathematics remains close to meaning, whereas school mathematics allows for a distancing from meaning and the acceptance of results that are unreasonable.

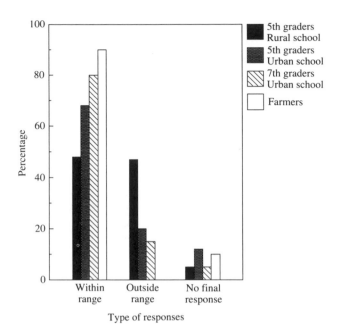

4.1   Percentage of responses for each group of S's inside/outside reasonable ranges in three problems with decimals. *Source for data:* Grando, 1988

This picture is strengthened by analyzing which relationships are assumed when arithmetic operations are chosen in the solution of a complex problem. This analysis is discussed in the next section.

b. How many trees on a piece of land?

Grando's problem on complex multiplicative relations (see Table 4.1) was viewed as involving a more elaborate problem situation, because subjects had to coordinate three multiplicative relations in sequence. In order to find out how many tea bushes would be planted on a piece of land 60 m by 40 m with spacing 3 m by 4 m between plants, subjects had to consider (*a*) the relationship between the length of the plot and one of the values in spacing, (*b*) the relationship between the width of the plot and the other value in spacing, and (*c*) the product of (*a*) and (*b*) (Strategy 1). Alternatively, subjects could consider (*a*) the total area of the plot, (*b*) the area used by each plant, and (*c*) the relationship between (*a*) and (*b*) (Strategy 2).

Table 4.3. *Explanations offered by a farmer of the steps he used in solving a problem*

Problem: Suppose this is a piece of land [shows a rectangle drawn to scale where the measurements 60 m x 30 m are indicated for length and width]. When you plant tea bushes, the space between the bushes has to be 3 by 4 meters. How many bushes will be planted in this plot?

FARMER: There will be fifteen bushes per row (of plants).

INTERVIEWER: And how do you know that it is fifteen per row?

FARMER: Because in each four meters you plant one bush. Then ten bushes will give you forty meters, but there are still twenty meters to plant. Then you need five more bushes. Its four times five, twenty. Then it is fifteen bushes per row.

INTERVIEWER: Right, fifteen per row.

FARMER: Then there are thirty meters on the side. Thirty by three. We'll see what else.

INTERVIEWER: I don't know what else.

FARMER: Well, thirty by three is ten rows. Ten rows in the front by fifteen on the side. That's one fifty. Then it's right, it makes one hundred fifty bushes.

Source: Data from Grando, 1988; translated from Portuguese by the authors.

These ways of approaching the problem were considered meaningful by Grando, who pointed out that subjects using these strategies could often say what they were trying to figure out at each step. For example, subjects using the first strategy above would indicate that they were trying to figure out how many rows of plants fit in the length of the plot when they divided 60 by 3. They would then indicate that they were trying to figure the number of plants per row when they divided 40 by 4. The final step in Strategy 1 was interpreted by subjects as figuring out the total number of plants in all rows. Similarly, most subjects using strategy 2 could explain at each step what they were attempting to figure out. One protocol by a farmer (using Strategy 1) is presented in Table 4.3 as illustration of this point.

This preservation of meaning in setting up the problem-solving strategy illustrates how a mathematical schema of a situation can represent both situational aspects and mathematical relationships. A schema of this sort is naturally somewhat general but also somewhat specific to similar situations.

Grando evaluated the proportion of meaningful strategies displayed by students and farmers in solving this problem. Overall results are summarized in Table 4.4. All farmers who gave final

Table 4.4. *Percentage of subjects by type of strategy*

Subjects	Meaningful/correct	Meaningful/incorrect
Fifth grade rural, ($n = 20$)	15	15
Fifth grade, urban ($n = 20$)	30	20
Seventh grade, urban ($n = 20$)	35	25
Farmers ($n = 12$)	83	0

answers to the problem used one of the two meaningful strategies described above, but fewer than half of the fifth graders and only 60% of the seventh graders used meaningful strategies.

Not all meaningful strategies used by students were correct. One type of error was produced by meaningful attempts that stopped halfway to the full solution. For example, some subjects said they had to calculate how many plants would fit along the length (or width) of the plot but did not know what to do next. The second meaningful but wrong strategy used by students consisted in attempting to follow Strategy 2 but calculating the area through addition instead of multiplication. Despite the error, these students were still able to state what they were trying to calculate, thus displaying their ability to link the schema of the situation with mathematical relations in problem solving.

Grando's study produced clear and provocative results, but it did not involve enough specific information on farmers' and students' use of mathematical knowledge to allow for the drawing of connections between their processes of acquisition of mathematical concepts and the models used in the experimental situation. Schliemann's (1984) study with carpenters and apprentices overcomes this difficulty by providing more information on the uses of mathematical knowledge by these two groups of subjects and predicting their solutions to a specific problem.

IV   Carpenters and apprentices: Arithmetic for the shop and for school

Schliemann (1984) interviewed 15 professional carpenters who had learned their profession while working as assistants to the owner of a shop – in most cases their own fathers. Naturalistic observation of the daily work of these professionals revealed that arithmetical problem solving often occurs when a customer brings the carpenter a drawing or a photo of a piece of furniture to be made. Carpenters have to calculate how much wood they need to buy, because this is a factor to consider in deciding how much they will charge for the finished product. From large shops they buy wood already cut into standard boards that they cut into the smaller pieces used in putting together the piece of furniture.

The group of carpentry apprentices was composed of 28 adolescents from poor backgrounds, aged 14 to 18, who simultaneously attended a three-year training program in carpentry and an associated regular school program. Apprentices had from four to nine years of school instruction in mathematics.

Naturalistic observation of the activities in the carpentry school revealed that carpentry apprentices have two quite distinct sources of experience with arithmetic problem solving. In the shop they start their practical training by performing simple tasks, such as cleaning and polishing, and only after a year of training (i.e., during the second year of apprenticeship) do they begin to build furniture. At this stage, they receive instructions for building each piece of furniture, plus a drawing and a list of all the parts needed, each part specified in terms of length, width, and thickness. The apprentices' job is to cut these parts from the available blocks of wood. It is only in the third year of apprenticeship that students take responsibility for making up lists of the parts required to build particular pieces of furniture. Instruction in mathematics is carried out entirely independently of this practical training. In their classes, apprentices solve arithmetic, geometry, and drawing problems and during the second year of apprenticeship learn how to calculate area and volume.

The task posed to carpenters and apprentices was to find out how much wood would be needed to build five beds like the one presented in a drawing (see Figure 4.2). The model specified the width, length, and thickness of pieces without repetition of the information when pieces were clearly corresponding. For example, for the sides of the frame, the length was indicated for one piece and the width for the other, and the thickness of the boards was indicated for only two out of the six boards. This reduced form of presenting information is sufficient to read the dimensions of all boards using complementary knowledge of symmetries in bed frames. This form of presentation of

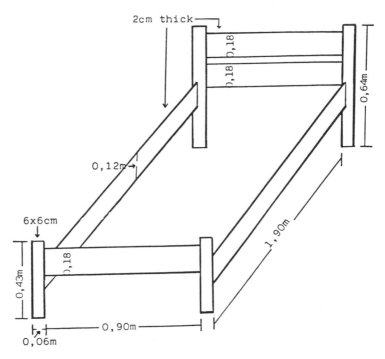

2cm thick

0,18

0,18

0,64m

0,12m

6x6cm

1,90m

0,43m

0,18

0,90m

0,06m

4.2 Drawing shown to carpenters and apprentices

the problem requires interpretation on the subjects' part from the outset. For example, by looking at the figure, subjects must realize that for the sides of the bed frame they need two boards with the dimensions 0.12 m by 1.90 m by 2 cm and that for the ends of the frame they need three boards measuring 0.18 m by 0.90 m by 2 cm. This information could be obtained by careful study of the drawing and was not directly conveyed to the subject.

Although the same problem was posed to carpenters and apprentices, they were not expected to respond to it in the same fashion on the basis of their experience. Carpenters' practice is to put together a list of boards that will be used in building specific models of furniture. Apprentices' practice is to cut out single boards from blocks of wood in a way that maximizes the use of the block. Thus, given the problem, carpenters are expected to make up a list of the parts and perhaps go on to find the total volume of wood used, which constitutes the basis for calculating the price. Apprentices are expected to think about blocks of wood that must be cut so that the desired boards are obtained. Further, apprentices have systematic exposure to mathematics classes in which they work in written fashion and where they have learned to calculate volume and calculate with decimals. Thus, apprentices are expected to show more written calculation, to follow algorithms when calculating with decimals, and to show greater loss of meaning as a consequence of these practices.

In summary, differences in the types of mathematical practice used by carpenters and apprentices should be noticeable both in the interpretation that is assigned to the problem and in the type of representation that is used in solving the problem.

As in the previous study, subjects were interviewed individually in a familiar environment: Carpenters were interviewed in the shops during working hours, and apprentices were interviewed in a classroom in their school. All problems were presented orally, but paper and pencil were provided. Subjects were told that they could use paper and pencil if they so wished, but were not asked to do so at any point in the interview. While they were trying to solve the problem, the examiner talked to them and discussed details of the drawing as well as the steps they followed in order to find a solution. Sessions were tape-recorded; an observer took notes, which were used together with tape transcripts and written material produced by the subjects in the analysis of their responses.

The analysis of results was carried out in three steps that aimed at (*a*) verifying whether the information was properly extracted from the figure, (*b*) identifying the subjects' interpretation of the task in terms of what they attempted to obtain as a result, and (*c*) identifying the practice of arithmetic called into play.

a. Extracting information from the drawing

Identifying the number of parts with the same dimensions did not represent a problem either for carpenters or for apprentices. All subjects had a clear understanding of which pieces had to be symmetrical and constructed lists of pieces to be used in building the bed using this knowledge. However, specifying the dimensions of the pieces proved to be a more difficult task. A piece is defined by three dimensions: length, width, and thickness. In the list, the use of these three dimensions to characterize the pieces to be used in making the bed varied across groups of subjects. Roughly one third of the

Table 4.5. *Number of subjects in each subgroup according to dimensions considered in trying to solve the problem*

Subgroup	Length	Length and width	Length, width, thickness
First-year apprentices	4	4	6
Second- and third-year apprentices	0	5	7
Professional carpenters	0	0	13

first-year apprentices took into account only the length of the pieces, and another third took into account only length and width, which are the most salient dimensions of a board. Only a third of the first-year apprentices took into account all three dimensions. None of the second- or third-year apprentices took into account one dimension only; about half of them considered both length and width, and the other half took all three dimensions into account. It will be recalled that first-year apprentices' experience in the shop involved only sanding and cleaning boards. Second- and third-year apprentices are engaged in cutting boards out of larger wooden blocks according to the appropriate length and width, the thickness being dealt with later by running the blocks through a machine. Finally, carpenters always identified pieces by taking all three dimensions into account. Table 4.5 shows the distribution of subjects according to the dimensions of the parts they took into account when trying to solve the problem. The correlation between number of dimensions considered and the level of mastery of carpentry was very significant (Kendall's tau = 0.46, $z = 4.18$, $p < .001$).

b. Interpreting the purpose of the task

Subjects extracted information from the drawing and used this information in ways that were clearly influenced by their everyday practice. It will be recalled that carpenters, upon receiving a drawing, make up lists of standard boards available in commerce in order to calculate how much wood they have to buy. In contrast, apprentices are involved in cutting boards from larger wooden blocks and do not deal directly with what standard boards are available. Their concern in everyday practice is the relationship between boards

4.3   List of materials needed to build five beds produced by a professional carpenter

to be cut and large blocks, which must be efficiently divided so as to avoid waste.

Strategies displayed in finding the final answers to the question "How much wood do you need to buy if you have to build five beds like the one in the drawing?" were classified into categories that reflect the interpretation the subjects gave the problem. These types of strategy are presented below.

List of standard pieces. Some subjects interpreted the question as a request for which and how many standard boards would have to be purchased if five beds like the one presented in the model were to be built. These subjects produced a list of standard boards from which all the pieces needed for the five beds could be cut. All professional carpenters and two apprentices displayed this strategy in solving the problem. Thus, carpenters' practice clearly influenced their interpretation of the question and determined their choice of strategy in problem solving. Figure 4.3 shows a list produced by a carpenter.

Computation of the dimensions of a single block of wood. A second group of subjects interpreted the problem as a request for the specification of a block of wood out of which all the pieces needed for the five beds could be cut. This interpretation was displayed in responses that consisted of specifying the length, the width, and (often but not always) the thickness of a single block of wood (that should be) equivalent to the total amount of wood needed for the five beds.

For an accurate definition of this block, subjects would have to consider successively the three types of boards defined by different lengths and widths and define the thickness of each of these three blocks by multiplying number of boards times thickness of the board.

For example, the boards used for the sides of the frame were 12 cm wide, and those used for the ends were 18 cm wide. Thus, separate calculations would be required for the sides and the ends of the frame. The sideboards would require a block of wood measuring 2 × 5 × [0.12 m (width) by 1.90 m (length) by 0.02 m (thickness)] – that is, number of boards in a bed (2) × number of beds (5) × [volume of a single board]. This block of wood needed for the sides would have to be *added* to the block needed for the ends and then *added* to the block needed for the poles in order to obtain an accurate specification of the block of wood with the total volume of wood needed for five beds.

However, subjects who specified one block of wood as their final answer did not proceed in this fashion. Instead, they added up separately the lengths, the widths, and the thicknesses of all the parts of the bed. The results of these three operations were then taken as length, width, and thickness of a huge block of wood from which parts should be cut. This procedure was used by 14% of the first-year and by 42% of the second- and third-year apprentices. Figure 4.4 shows the written material produced by a second-year apprentice who chose this wrong procedure. The answer he obtained, as a result of a wrong method and arithmetical mistakes, was a block measuring 16.38 meters in length, 10.20 meters in width, and 0.12 meters in thickness. This, according to his method, would be the amount of wood needed to build only one bed. This kind of wrong solution, accepted without criticism by the apprentices, never appeared among professional carpenters.

Addition of all dimensions considered. In this kind of answer, subjects obtained a single total of all the values representing the lengths, the widths, and the thicknesses (when all the dimensions were considered) of all parts of the bed. The result thus obtained was inadequately given as the number of either meters or square meters or cubic meters needed to build the beds. None of the professional carpenters presented this kind of solution, but 79% of the first-year and 50% of the second- and third-year apprentices did so. Figure 4.5 presents written material produced by two first-year apprentices who gave this sort of answer. In the first example, length was the only dimension considered; in the second, the three dimensions were added together.

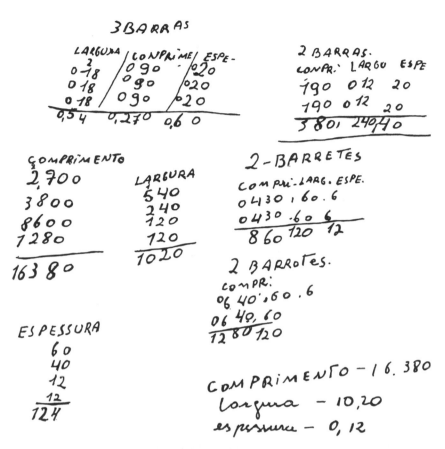

4.4   Written material produced by a second-year apprentice trying to solve the problem.

These three types of strategy are not equivalent in nature. Strategy 1, the list of standard parts, is a meaningful response. The other two involve a loss of meaning. Strategy 2 – obtaining the dimensions of a large block of wood from which all parts could be cut – would have been a meaningful and rather complex strategy. However, the strategy of calculating the dimensions of a single block of wood, which was used by students, was not carried out appropriately. It mixed pieces of different dimensions in the calculation of a single block and involved multiplications of three lengths, three widths, and three thicknesses when all dimensions were considered, resulting in a very distorted overall block. Finally, Strategy 3, the addition of all dimensions into a

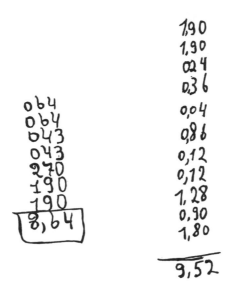

4.5   Written material produced by two first-year apprentices trying to solve the problem

single total, also involves loss of meaning. It is possible to argue that students using this strategy were trying to calculate the total volume of wood, which would yield an answer represented by a single number, but that they wrongly calculated volume additively instead of multiplicatively. Substitution of addition for multiplication has already been reported in the calculation of area (Grando, 1988) among Brazilian students. Volume would, in fact, have been a meaningful response to the problem, because it would represent a correct answer to the question "How much wood is needed to make the five beds?" Volume calculation was in fact carried out by four carpenters as an intermediary step in order to calculate the price of the wood to be used for the five beds. In shops, wood is sold by the standard piece, but prices are computed by volume. The solutions presented by these four professional carpenters, however, were correct and consisted in finding first the volume of pieces with the same dimensions and then adding the results for all the pieces. Apprentices, however, mixed calculations across pieces and used the wrong inputs for the calculation of volume.

Table 4.6. *Number of subjects in each subgroup according to the final answer given*

Subgroups	Addition of dimensions	Block from adding up each dimension	List of standard parts
First-year apprentices	11	2	1
Second- and third-year apprentices	6	5	1
Professional carpenters	0	0	13

The frequencies with which subjects in each group used each type of strategy are shown in Table 4.6. The correlation between the degree of mastery of carpentry and the kind of strategy used was calculated assuming an ordering from weakest (addition of dimensions) to best strategy (list of standard pieces). This correlation was very high (Kendall's tau = 0.73) and significant ($z = 6.50$, $p < .0001$).

In short, the analysis of strategies used in interpreting the question "How much wood is needed to make the five beds?" varied with the subjects' experience. Carpenters were able to use their experience to represent the problem in terms of which standard pieces they would have to buy to make the beds. Apprentices, whose experience was mostly with blocks of wood being cut into boards and with volume calculation in the classroom, interpreted the question as a request to define a large block of wood from which all parts could be cut. This block could be described either by its dimensions or by its volume. However, a considerable loss of meaning was observed during problem solving, and dimensions from different pieces were mixed improperly in the attempt to define the large block of wood.

c. Types of arithmetic practice used by carpenters and apprentices

The final analysis to be carried out relates to the identification of the arithmetic practice used by each group of subjects. We had hypothesized that different arithmetic practices, oral and written, would be connected to learning in the shop (carpenters) versus learning in the classroom (apprentices). We had also hypothesized that oral practices would remain close to meaning and that the loss of meaning observed among apprentices would be related to their use of mostly written arithmetic.

In order to verify these hypotheses, the procedures used by each subject in solving the problem were classified into three categories: (*a*) mental computation, when all computations were performed without using paper and pencil; (*b*) school algorithms, when paper and pencil were used to perform all computations in the usual school-type procedure, going from units to tens and then to hundreds; and (*c*) a mixed strategy, characterized by the use of mental computation for the operations involving smaller numbers and school algorithms for those involving larger ones.

Oral and written procedures were clearly similar to those described in previous chapters. Mixed procedures typically consisted of oral solutions permeated by isolated computations carried out in writing as a consequence of the difficulty of the particular computation. When the result was obtained, the subject immediately interpreted it in oral fashion and continued operating according to oral arithmetic. An example of a mixed procedure is presented below, taken from a protocol of a professional carpenter who had attended school for three years.

> In the drawing there is a pole with sixty-four [referring to the length of one of the poles in centimeters], another one, and two more with forty-three. Sixty-four [pause], one meter and twenty-eight [which is the result of the mental computation where 64 is considered twice]. And here [pointing to the shorter poles at the end of the bed] forty-three, right? That makes one meter twenty-eight plus eighty-six [that is, two poles of 43. The subject then writes 128 aligned vertically with 86 and calculates] four [writes 4 below the units column], carry the one. Three plus eight, eleven [writes 1 under the tens column], carry the one [then he writes 2 under the hundreds column]. OK, I need a piece with two meters and twenty to cut to measure [immediately recovers the meaning of the calculation, reading 2 meters and 20 to cut to measure, a reading which allows for losses, because the exact result was 2 meters and 14].

Table 4.7 shows the distribution of subjects by type of procedure used. It is clear that mental computation was the preferred procedure among professionals: 12 out of 13 professionals (92%) used mental

Table 4.7. *Number of subjects in each subgroup according to strategy used to solve arithmetical operations*

Subgroup	Mental computation	Mixed strategy	School algorithms
First-year apprentices	0	5	9
Second- and third-year apprentices	0	6	6
Professional carpenters	3	9	1

Note: Only one apprentice did not attempt to perform the task. Two professional carpenters who had never been to school gave a final answer without explaining how it was obtained. These three cases were not included in our analysis, thus reducing the number of subjects to 26 apprentices and 13 professional carpenters.

computation in isolation or combined with school algorithms. In contrast, almost twice as many first-year apprentices, whose experience was mostly related to school calculation, preferred a written to an oral strategy. Second- and third-year students' procedures were distributed equally between written and mixed strategies. This distribution of preferred procedures differs significantly from that which would be obtained by chance ($\chi^2 = 7.01$, $N = 39$, $df = 1$, $p < .01$). In some cases, the use of written school algorithms by apprentices occurred even when they had to add or multiply small numbers such as $6 + 6$ or 2×5.

Despite the differences in the preferred computation strategy, carpenters and apprentices did not differ in their ability to compute. Errors in the computations per se were rare: Of the 292 operations performed by all subjects, only 31 were wrong. Thus, computing strategies, although different, were equally effective in both groups. This is a finding of interest if one bears in mind that levels of formal school attendance were very different both across the two groups of and within the group of carpenters. It is also worth mentioning that more economical strategies, such as using multiplication instead of addition, were significantly more frequent among carpenters than among students. These results are in keeping with those described by Scribner (1984 a,c), who observed that the use of what she called *optimizing strategies* increases with professional expertise and practice rather than with schooling.

V Conclusions

Taken together, the studies described above show that mathematical problem solving involves the use of two types of representation: (*a*) representation of the problem situations and (*b*) representation of mathematical relations. A good problem solver must be able to connect the two types of representation quite easily, pulling out the mathematical relations from a problem but also turning the mathematical relations around in ways not suggested by the meanings in the problem situation.

The two studies in this chapter show that mathematical knowledge, despite its abstract nature, seems to bear the marks of the social situations in which it was acquired through the type of symbolic representation used in these different situations. School mathematics is learned mostly as written mathematics. Representations of the situation are abandoned as much as possible for the sake of generality. In consequence, mathematical relations represented in school mathematics have poor ties with problem situations. Students ask themselves which operations to use and find difficulty in interpreting the meaning of their answers after calculation. Street mathematics is practiced mostly in the oral mode. Oral representation preserves meaning and consequently helps problem solvers to work from the meaning of problem situations to mathematical relations. A question like which operation to use and difficulties in the interpretation of responses are unusual.

The comparisons between farmers and students show that unschooled or barely literate farmers can perform better than students with five years of schooling in solving simple arithmetic problems. Their problem-solving strategies are closely related to the meaning of the problem situation, and as a rule, their answers fall within reasonable ranges.

The comparisons between carpenters and apprentices also show that mathematics learned in everyday life may result in a better performance in problem solving than school learning. Carpenters and apprentices used strategies closely related to the social situations in which they had learned mathematics and to the tasks they had to carry out. Whereas professional carpenters attempted to find a list of standard pieces that they could then purchase in order to build beds, most apprentices tried to find the measures of a block of wood out of which they could cut boards for the beds. Both carpenters' and apprentices' strategies reflected their different tasks in the social contexts in which their mathematical knowledge was developed.

The results of these two studies are not simply an expansion of the previous findings about children's oral and written arithmetic. The children interviewed in the previous studies were younger and had less schooling. Their difficulties in written arithmetic could be traced directly to their difficulties with school algorithms. They showed error patterns traditionally described in the literature about addition and subtraction algorithms, such as failure to borrow, subtracting the smaller from the larger, and the "zero error" (see, for example, Brown & Burton, 1978; Carraher & Schliemann, 1985; Young & O'Shea, 1981). The students who took part in the two studies described in this chapter all had higher levels of schooling and were no longer prone to these traditional computational errors. They could carry and borrow across columns in written arithmetic. In fact, practically no computational errors of this nature were observed among apprentices in the carpentry shop. The difficulties observed among students in the two studies described in this chapter were connected to their concentrating on the numbers rather on the meaning of the problem. The gains in performance resulting from learning mathematics outside school could be traced to the preservation of meaning in choosing problem-solving strategies and interpreting answers.

A final word about the problems used in these studies is, however, necessary. The solutions to problems in these studies could in fact be modeled mentally from the meanings of the situations. All problems proceeded in the same way practice usually does. For example, if a farmer wants to plant tea bushes, it will be necessary to find out what the spacing between the bushes ought to be and then calculate how many bushes will be needed in order to fully plant the area to be used for cultivating the bushes. This is the type of question that was posed in these studies. But what would farmers do if we posed the inverse problem? In other words, what would happen if we told them the size of an area and the number of tea bushes in it and asked them to figure out the distance between the bushes? Similarly, carpenters work from orders of pieces of furniture to figuring out how much wood they need. But how would they approach the inverse problem; that is, finding out how many beds could be built with a certain amount of wood? Inverse problems cannot be so simply modeled from the meaning of the problem situation. They seem to require further knowledge about mathematical relations – such as which operations are the inverse of one another. In the chapters discussed so far we have not dealt with inversion. In the two chapters that follow, inversion will become the focus of our analysis.

5 Situational and mathematical relations: A study on understanding proportions

I Introduction

Studying the development of mathematical concepts in and out of school requires careful analysis of the interplay between schooling and other cultural practices in which specific mathematical concepts may be acquired. A clear description of cultural practices, regardless of whether they are carried out in school or outside, is needed if we are to see the connections between the learning situation and later problem-solving ability. In chapter 4 we raised the possibility that out-of-school practices may contain so many situational elements in the representation of problem-solving strategies that the strategies may be limited in application. Practice that is unidirectional may result in a lack of reversibility; that is, difficulty in solving problems the direction of which is the opposite of one's everyday practice. In contrast, school learning, which includes fewer situational elements, may be less prone to this lack of reversibility and thus more flexible in its application.

This chapter focuses on the understanding of proportions in and out of school and on the ability to solve inverse problems. The next section briefly presents a psychological analysis of concepts and problems involving proportions. The third section looks at some cultural practices related to the understanding of proportions, describing

We are grateful to the *International Journal of Behavioral Development* for permission to reproduce Tables 5.1–5.3 and other substantial portions of the paper "From drawings to buildings," which appeared in vol. 9 (1986), 527–544. We also thank the people and institutions that made this work possible. Our friends Alvimar Moreira da Silva and Laércio Campos gave us permission to spend much time observing foremen at work and to use blueprints in the interviews. Lourdes Carneiro Leão, Sônia Proto, and Ana Carolina Brandão helped with data collection. CNPq and MEC/CAPES/SPEC/PADCT supported this study with grants.

how proportions are taught in school and one situation in which proportions may be learned outside school. Finally, in section IV, a study is described that contrasts the understanding of proportions developed by students in school and by foremen working on construction sites, whose knowledge of proportions was most likely developed outside school.

II A psychological analysis of the concept of proportions

Piaget's work has undoubtedly made a great contribution to current views about the nature and development of several logicomathematical concepts, including the concept of proportionality. Piaget and his collaborators (see Inhelder & Piaget, 1958; Piaget, Grize, Szeminska, & Bang, 1968) studied the development of the concept of proportionality using several different problem situations, such as the relationship between the length of an eel and the amount of food it had to be fed, the understanding of probability, the relationship between weights and distance from the fulcrum in a balance scale, and the understanding of the relationship between the size of a shadow and the distance between the object and the source of light, to name just a few examples. Throughout these studies, Piaget and his collaborators searched for a general description of the development of the concept of proportionality and proposed the following stages. In a first stage, children seemed either to focus on one variable only or to take no account of the relationship between the variables. In a second stage (IIA), children started noticing this relationship, but only in an intuitive fashion. For example, they may have noticed that if a increases, so does b, but in a prediction task they would make their predictions without any systematic effort to quantify the increases in a and b. At the next stage (IIB), children started to quantify increases in the variables in more systematic ways but did so additively and not by keeping a constant ratio between the two variables. For example, if a was increased by 4 units, children would expect that b should also be increased by 4 units. This was considered by Piaget and his collaborators a significant improvement in comparison with the previous stage, because children at this level took into account the two variables and related them systematically, although using an incorrect additive relation. Finally, in a last stage (III), that of formal operational thinking, youngsters would take both variables into account and relate them by a constant ratio.

Piagetian theory leads to the expectation that the particular problem situation does not matter very much. Once children are able to understand the concept of linear function, they should be able to do so in a variety of situations and with a variety of ratios. Although the idea of variation in the ability to solve problems with the same logicomathematical structure was accepted during the concrete operational stage, coming under the notion of *horizontal décalage*, this variation was not expected to be significant during the formal operational stage. However, research on the concept of proportions has shown that children may perform differently depending on the type of response they have to make (that is, prediction of a value or evaluation of the equivalence of two ratios; see, for example, Karplus, Pulos, & Stage, 1983), the numbers involved in the problem (whether the ratio is an integer or a fraction; see Hart, 1984, and Karplus, Pulos, & Stage, 1983), and the type of problem situation (see Vergnaud, 1983). Thus, Piagetian theory on the understanding of proportionality needs a complementary analysis of situational models that embed the same logicomathematical invariants.

Vergnaud (1983) has suggested that three different situational models can be distinguished as underlying problems in proportion; (*a*) the isomorphism of measures, (*b*) the product of measures, and (*c*) the multiple proportions model. These three models are described succinctly below.

In the *isomorphism of measures* model, only two variables are involved, and each value on one variable corresponds to one value on the other variable. This is essentially a one-to-many correspondence situation: To the value 1 in variable *A* corresponds value *x* in variable *B* and, accordingly, to *n*(1) corresponds *n*(*x*). For example, the relationship between number of items and price in a problem situation can be described by the isomorphism of measures model.

In the *product of measures* model, (at least) three variables are involved, the third variable being the product of the first two. For example, the area of a rectangle (variable 3, measured in square meters) is the product of the length (variable 1, measured in meters) and the width (variable 2, measured in meters). Because more than two variables are involved, this type of situation cannot be described by the simple one-to-many correspondence schema of the preceding

Isomorphism of measures

Product of measures

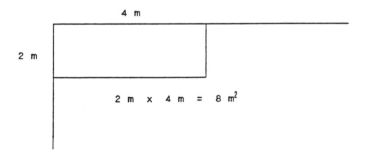

5.1 In the isomorphism of measures model, variables (price and weight) change in corresponding fashion. In the product of measures model, two measures (length and width) give rise to a third (area)

model. There are at least double correspondences, which can best be represented through a Cartesian product table. These two models are schematically represented in Figure 5.1.

The *multiple proportions model* corresponds to situations that are essentially similar to the isomorphism of measures model but in which several variables are involved. For example, the income a farmer can get from milk on his farm is (in a general fashion) proportional to the number of cows he has, the number of days over which production is considered, and the price of milk. Let us suppose that one cow (variable 1) produces on average 30 liters of milk (variable 2) per day (variable 3) and that each liter of milk costs $0.50 (variable 4). It is possible, then, for the farmer to calculate approximately how much income he will take from milk production in a certain period of time. When all other variables are held constant, there is a simple correspondence between the number of cows and the amount of milk produced. The same is true for the correspondence between the number of days and the amount of milk produced or income obtained. Figure 5.2 gives a schematic representation of the multiple proportions model.

Our investigations will focus only on the situations described by the isomorphism of measures model. This model was chosen because, according to Vergnaud, people's initial understanding of proportion is based on the isomorphism of measures model and does not immediately apply to all other types of situations. The isomorphism of measures model is simpler for many reasons. The first – and most obvious – reason is that fewer variables are involved. The second reason is perhaps more important. In the isomorphism of measures model, *the understanding of the situation is kept in view when the mathematical relations are analyzed. Each variable remains independent of the other in the subject's conceptualization, and parallel transformations are carried out on both variables, thus keeping their values proportional.* For example, if one sells/buys something, there is usually a price *x* that is set in correspondence with each unit of the product sold. When there is an increase in the number of units, there is a proportional increase in the price – that is to say, as many *x*'s are added to the cost as items are increased in the purchase. The psychological reality of this description is instantiated in some of the protocols in chapters 2 and 3 describing occasions when children explicitly kept separate accounts of the number of items and the total price. For example, M., calculating the price of 10 coconuts, said:

> Three will be one hundred and five. With three more, that will be two hundred and ten. I need four more. That is . . . three hundred and fifteen [price of nine coconuts]. I think it is three hundred and fifty. (See chapter 2, sect. III.2.a)

No. of cows	Average prod p/ cow	No of days	Price p/ liter	Income
1	30	1	.50	15.00
x5			x 5	
5				75.00
		x 30		x 30
		30		2,250
			x 1.2	x 1.2
			.60	2,700
x 1.1			x 1.1	
				2,970

5.2 Schema of multiple proportions. Several variables affect income, and thus there are several correspondences

Thus, it seems possible to speak of a psychological schema of isomorphism of measures developed as a model for empirical situations. *This schema includes the situational meaning and the mathematical relations in the same representation.* Since the situational and mathematical relations are represented by one schema only, it is possible that the schema of isomorphism of measures will remain closely connected to the cultural practices in which it is developed and will incorporate some of the particularities of the situation. For example, in the everyday practice of calculating prices, one usually knows the price of one unit (or a set with a fixed number of units) and calculates the price of any *larger* number of units. That is to say, one knows the pair of values $[1 : x]$ and can *go up* to any other pair $[n(1) : n(x)]$. Practice tends to be unidirectional. Even subjects who know well procedures for solving problems in a unidirectional practice may not be able to invert the procedure. Further, the operation of addition is sufficient for the type of problem in which one calculates the price of larger amounts on the basis of smaller amounts. For example, 1 item costs x; 2 items cost $(x + x)$; 4 items cost $(x + x) + (x + x)$; 5 items cost $(x + x) + (x + x) + x$, and so on. However, the inversion of the mathematical relations in this problem cannot be accomplished by subtraction, which is the inverse operation of addition. Inversion here requires division (conceived as the inverse of multiplication and not in its original sense of sharing). Solving the increase in items through addition may obscure the multiplicative nature of the situation. Inversion requires the recognition that multiplication is involved in the problem and that its inverse is division. If it turns out that the schema of isomorphism of measures developed in everyday life does remain closely connected to the cultural practices from which it emerges, street mathematics will not be flexible. Inversion will be difficult, because cultural practices tend to be unidirectional. Transfer from one isomorphism of measures situation to another should also be difficult as a result of the interconnectedness of mathematical and situational relations. In other words, if situational information and mathematical relations are inextricably interconnected in the model, it becomes impossible to envisage a general mathematical model that can be applied to other situations. Support for this hypothesis would substantiate the claim that everyday mathematics (or everyday knowledge in general) is less abstract than school mathematics (or scientific knowledge).

On the other hand, if people who have learned their mathematics outside school turn out to be able to solve inverse problems and transfer proportional reasoning to new situations, there will be less support for the idea that everyday mathematics is inherently different from school mathematics. Although a schema of isomorphism of measures may contain both situational and mathematical relations, it is still possible that people can reflect upon the mathematical relations understood in context and develop a more general mathematical understanding of relationships between variables.

We now turn to a description of the learning environments of students and construction foremen in Recife in order to describe the kinds of learning situations they were exposed to with respect to proportions. In section IV, we look at how their knowledge was used in solving a proportions problem.

III   Foremen and students: Social and learning contrasts

a. Foremen

In the Brazilian class-differentiated society, an urban lower class can be identified that encompasses a whole range of occupations. These occupations include (*a*) the informal sector of the economy, described briefly in chapter 2, and (*b*) regular jobs in industries. Some of these are unskilled; others require the development of semispecialized skills, the latter affording higher status in the work environment as well as greater job stability, better chances of employment, and levels of pay above the minimum wage. Construction foreman is an occupation that falls into the latter group. Construction foremen usually receive their training informally through an apprenticeship with a family member or close friend. Even though specialized training has occasionally been available in the cities in the past two decades, training opportunities are scarce and do not influence the job market. The foremen interviewed in the study described in this chapter became skilled in their jobs over a period of one or more years of supervised practice; only 1 (out of 17) had had (six months of) formal training.

Mathematics is an integral part of a foreman's work, which involves frequent measurement and calculations. Among many other responsibilities, foremen have to calculate the amounts of building materials delivered to construction sites (e.g., bricks and sand), calculate the exact amount of work done by bricklayers in terms of square meters, and do the necessary measurements to set up guidelines to demarcate external and internal walls of buildings, making sure that length, width, and angles are correct according to the blueprints. They work from blueprints drafted with such precision that it is possible to determine the exact real-life measurement of a wall from the blueprint. This precise form of drawing is called scale drawing. Three or four scales are commonly used by foremen in Recife: 1 : 100 (in which 1 cm in the drawing represents 100 cm in reality), 1 : 50, 1 : 20, and 1 : 250. A blueprint always displays the scale used. Further, real-life dimensions of the walls are also usually identified on the drawing, although occasionally the length of a wall or the width of a hall, for example, may be omitted. In this case, the foreman can measure the wall on the blueprint and use the scale in the drawing to calculate the real-life length of the wall. In the 1 : 100 scale, a wall drawn as 3 cm will be 3 m long in reality. If the scale were 1 : 50, 6 cm would stand for 3 m. From a known scale and a known dimension in the drawing, the foreman can figure out the unknown real-life dimension.

b. Students

Youngsters growing up in the urban lower class are often exposed to building activities carried out at home by their fathers, for much of the housing for these families will be constructed by family members. In order to insure that students in the present study had not been exposed to this type of learning environment, students were sampled randomly from a class of seventh graders in a school that serves a middle- and upper-middle-class clientele. As part of their seventh-grade math curriculum, they received instruction on the "rule of three," the algorithm taught in Brazil for solving proportion problems. The rule of three prescribes that when one knows three values and wants to find the fourth, the numbers are set up in this format:

$$\frac{a}{b} = \frac{c}{x}.$$

This equation is then solved by cross multiplication, thus indicated:

$$\frac{a}{b} \diagdown\hspace{-0.6em}\diagup \frac{c}{x},$$

The equation is then written as

$$a \times x = b \times c,$$

and thus

$$x = (b \times c)/a.$$

The only aspect of meaning taken into account is whether values in the variables involved increase together or whether one increases as the other decreases. Children are taught that if values in one variable increase as the values in the other variable also increase, then proportions are direct and the model is applied in a straightforward fashion. If, however, the values in one variable increase while the values in the other variable decrease, x and c will be shifted around (x goes on top and c on the bottom) before the procedure is applied. This procedure is very general and can also be applied in a step-by-step fashion to multiple proportions problems. If correctly used in proportions problems, it yields 100% correct responses. No explanation is usually given to students for the shifting around of the numbers in inverse proportions problems, and this manipulation seems to be quickly forgotten (see Carraher, Carraher, & Schliemann, 1985). Little attention is given in math textbooks to connecting the mathematics with the problem situation, and the initial phases of teaching involve mostly formal demonstrations. The formal demonstrations are followed by exercises in application of the procedure. *In the applications, it is assumed that the procedure just learned is appropriate; therefore students do not concentrate on a discussion of what connections there may be between mathematical models and empirical situations.* This type of instruction can be characterized as concentrating solely on the mathematical relations despite the often numerous application exercises at the ends of chapters, because the only situational consideration necessary for solving the exercises successfully is whether the values increase together or whether values on one variable decrease as those in the other increase.

Problem situations created in the context of scale drawing are not familiar to most students; no problems of this nature were found in the textbook used by the students who took part in this study. Although students know what meters and centimeters refer to, the meaning of these measurements may not be wholly clear to them. The distance between their school knowledge of meters and centimeters and their everyday notions about space was well illustrated in a study by D. W. Carraher (1985), who asked fifth- to eighth-grade students to draw their houses and estimate the size of each room. He obtained such responses as "one meter by ten centimeters" for the estimated dimensions of the students' own bedrooms!

The diversity of experience in these two groups, who share the same measurement system – the metric system – and many other experiences involving number (such as the same system for writing numbers and the same monetary system), creates an interesting situation in which one can contrast the impact of schooling with the impact of work upon the development of mathematical knowledge. Foremen seemed to have acquired their mathematical knowledge mostly outside school, learning in situations that provided meaning for mathematical representations. In contrast, students, although living in a culture rich with situations that could be used in the classroom to make mathematical representations meaningful, were taught mathematics in a way that distanced it from everyday meanings. How would these two groups of people perform in a task that was somewhat novel to them?

IV   Solving problems about proportions

T. N. Carraher (1986) interviewed 17 foremen and 16 students randomly drawn from a seventh-grade class in a private school in Recife. The students ranged in age from 13 to 16 and had all received instruction on the rule of three. The foremen ranged in schooling from illiterate to high-school level; only three had reached seventh grade, the grade level at which the rule of three is usually taught. All foremen had at least five years of practice in the trade. They were located through visits to construction sites in three neighborhoods. No selection of a sample with particular characteristics (such as previously specified levels of schooling or practice) was attempted.

a. Method

Subjects were shown four blueprints, one a time, drawn to different scales. The drawings, however, did not specify the scale used, as is usual on blueprints. When each blueprint was shown, it was pointed out to the subject that for most of the walls, the

Table 5.1. *Problems presented in the study*

Scale	First pair	Known values in second pair
1 : 100	3 cm/3 m	4 cm; 2.8 cm; 3.2 cm
1 : 50	6 cm/3 m	9 cm; 7.5 cm; 5 cm
1 : 40	5 cm/2 m	8 cm
1 : 33.3	9 cm/3 m	15 cm

Note: Known values in the second pair were obtained by the subject through measurement; some slight variation was observed as a consequence, e.g., 3.3 instead of 3.2 cm.

measurements were indicated on the blueprint. However, for some of the walls there were no measurements. Their task was to figure out what those measures were by using information from the blueprint. As a start, they obtained one measure from the blueprint and compared it with the measure indicated on the blueprint for that same wall; these two values were the first pair of numbers. This first pair represented necessary information for the application of the rule of three and for determining the scale. The second pair of values was a pair for which there is an unknown, the real-life measure, and one known measure obtained from the blueprint. Table 5.1 presents a summary of the numerical problems used in the study. Figure 5.3 is a blueprint drawing to scale 1 : 40.

This task was new to both foremen and students, but in different ways. For foremen, the task involves a familiar content but requires inversion of their everyday practice. At work they may have to calculate the real-life size of a wall, but the scale used in the drawing will be known – it will be written on the blueprint. In this task, the scales are unknown and have to be determined from a pair of known values obtained from the blueprint. For students, the task is novel because of its content: They had not worked out proportions problems about scale drawing as part of their school tasks. However, schoolbooks carry problems in different forms and without a constant direction for the calculations. Thus, we have no reason to expect the students to have problems with inversion.

Drawings in four scales were used, two of which (1 : 100 and 1 : 50) are commonly used by foremen, and two of which (1 : 40 and 1 : 33.3) are not used at all. The scales 1 : 100 and 1 : 50 are there-

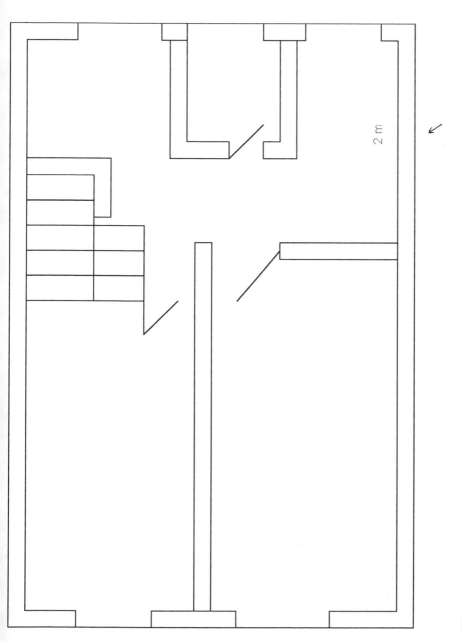

2 m

5.3 Drawing used in the problem for scale 1 : 40

fore familiar to foremen, and the scales 1 : 40 and 1 : 33.3 are unfamiliar. The distinction between familiar and unfamiliar scales does not apply to students, who had no specific contact with scales but only general instruction on proportions problems. The two sets of familiar and unfamiliar scales were presented to all subjects in a fixed order, the familiar scales being presented first. Within each set, order of presentation was systematically alternated.

Subjects were interviewed individually according to the Piagetian clinical method. Additional questions were asked after the initial presentation of the items in order to identify the strategy used in solving the problem when no information had been obtained through intermediary calculations carried out in an observable fashion. Paper and pencil were available, but there was no attempt to influence subjects to work out solutions in writing. A ruler was also available and was used by the interviewer to obtain the values in the initial problem. The interviewer thus created a model for how to obtain the information needed to solve the subsequent problems.

b. Results

Three aspects of performance were analyzed: accuracy of solutions, strategies in problem solving, and types of errors. These are, of course, not independent aspects of performance, but their analysis in greater detail helps us to build a clear picture of the influence of different types of practice on problem solving.

Accuracy. The percentages of correct responses for each scale observed among foremen and students are presented in Figure 5.4. Foremen did significantly better than students on both familiar scales (for the scale 1 : 100, $z = 3.8$; $p < .01$; for the scale 1 : 50, $z = 4.22$; $p < .01$), but there was no significant difference between the two groups on the unfamiliar scales. Looking only at the results observed among foremen, the overall percentage of correct responses on the two problems with familiar scales was compared with the overall percentage of correct responses on the two problems with unfamiliar scales. This analysis indicated a significant effect of familiarity (according to a test for correlated proportions, $z = 2.875$; $p < .02$). The effect of familiarity clarifies why foremen did better than students on familiar but not on unfamiliar scales. Foremen's perfor-

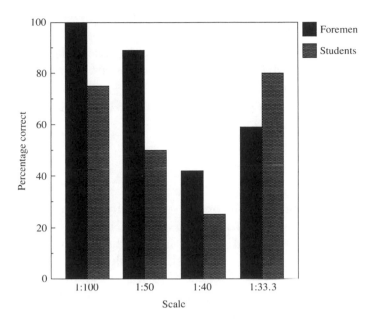

5.4   Percentage of correct responses for each scale

mance showed a significant drop in problems with unfamiliar scales, from performing almost at ceiling level in the familiar scale to about half correct responses in the unfamiliar scales. Thus, familiarity with the scales influenced the foremen's ability to solve inverse problems, and no simple conclusion can be drawn about the reversibility of their knowledge on the basis of this quantitative analysis. The qualitative analysis that follows will shed some light on the reasons for the drop in performance on unfamiliar scales among foremen.

Strategies used in problem solving. The strategies used in problem solving were classified according to the categories indicated below. This classification was based on the calculations carried out in observable fashion during problem solving and the subjects' explanations for their responses. Some of the categories used were chosen for theoretical reasons, and some were empirically derived.

The theoretically interesting categories were the *rule of three*, which is the algorithm taught in school for proportions problems, and *incorrect additive solutions*, which are described by Piaget and his coworkers as characteristic of the late concrete operational period (IIB).

It could reasonably be expected that both kinds of response would be observed in the study. The two remaining categories are essentially descriptive and were used to characterize the observed performances. They were labeled *hypothesis testing* and *finding the relation*. These strategies are briefly described below.

The *rule of three* strategy involves applying the algorithm taught in school for solving proportions problems:

$$\frac{a}{b} = \frac{x}{c}$$

The rule of three could be applied successfully to any problem in this study: Subjects had data on the first pair of numbers (a/b) and one known value in the second pair. It is noteworthy that the computations necessary for applying the rule of three (multiplication and division) are learned in school in second or third grade, and that students' failure to apply the algorithm is much more likely to derive from conceptual difficulties than from the specific computations required.

Incorrect additive solutions consist of taking the absolute difference between the values in the first pair of numbers and making this difference constant in the second pair. For example, the first pair of numbers in the 1 : 50 scale was 6 cm in the drawing standing for 3 m; the difference between the two values is 3. A wrong additive solution would be, for example, the answer "6 m" when the measurement obtained from the blueprint was 9 cm. As Piaget and his co-workers point out, this type of response takes into account a relationship between the two variables in the problem but does so in an additive and not a multiplicative way.

Hypothesis testing was a strategy observed in this study according to which the subject seemed to treat all familiar scales as a pool of hypotheses. Each of the hypotheses in the pool could be checked against the data available in the first pair of numbers. This would lead the subject either to accept the hypothesis because the data confirmed its validity or to reject it because it generated predictions inconsistent with the data obtained from the blueprint. After the scale has already been identified, the subject proceeds to solve the problem of calculating the real-life dimension of the target wall by calling upon his usual practice. The strategy thus involves two steps: (*a*) identifying the scale on the basis of the first pair of numbers and (*b*) calculating the unknown in the second pair. The transcripts from the protocols below show how subjects take the information from the first pair of numbers and check it against the values expected for the scales they are familiar with. Two examples are given, one of a successful solution with the familiar scale 1 : 50, and the second of a failure with the unfamiliar scale 1 : 33.3. When the scale is familiar, the first step appears to be based on well-memorized information.

> J. A., for example, obtained the information on the first pair for the 1 : 50 scale.
>
> J. A.: This is not the one by one hundred scale, this is the one by fifty because in the scale one by one hundred each centimeter counts as a meter. Here you have to make it less [i.e., 6 on the blueprint represents 3 in real life]; then you know it is not the scale one by one hundred.

When the scale was unfamiliar, it did not belong to the subject's pool of hypotheses. Thus, after testing for all the hypotheses in the pool, the subject could not proceed. Consequently, this strategy was not effective with unfamiliar scales.

> Working with the unfamiliar scale 1 : 33.3, L. S. checks the first pair of numbers against the values expected if the scale had been 1 : 50, speaking out loud. After some pauses, he rejects the hypothesis that the scale is either 1 : 100 or 1 : 20, which are the other two scales he knows from his work.
>
> L. S.: Nine centimeters, three meters. This scale is . . . one by fifty, no, that would be four meters and a half. [Pause] If you drew it like this, that is because it is correct. [Pause] Can't do it.
>
> E.: Why not? You did solve all the others.
>
> L. S.: Because it doesn't work for one by fifty, it doesn't work for one by one [meaning: 1 : 100] and it doesn't work for one by twenty. There are three types of scale, one by fifty, one by twenty, and one by one. The simplest scale is one by one, you don't have to work on it, you look at the centimeters and you know the meters. Now, one by fifty and one by twenty you have to calculate. Now, this one here, it shows nine centimeters by three meters. I've only worked with the other three.

This strategy is effective both on the job, where all scales are written on blueprints, and outside work situations in solving inverse problems with familiar scales. However, it does not allow for generalization to solving inverse problems with other scales. The strategy seems to involve the representation of the relationships at a very specific level – in scale 1 : 100, 1 cm counts as 1 meter; in scale 1 : 50, 1 cm counts as 50 cm. Foremen using this strategy accomplish the solution of inverse problems without actually inverting the mental operations. When they take a known scale as a hypothesis, they think about the problem in the usual way they do on the job. If the hypothesis is supported by the data, they continue to a solution. If it is not, they take one of the remaining scales as a hypothesis or give up (if no other scale remains to be tested). It is clear that hypothesis testing is a strategy that can be effective only when the range of possible solutions is restricted. In a problem with a very large (or infinite) number of possibilities, the chances of coming up with the correct hypothesis are much reduced.

Finding the relation is a descriptive category developed in view of the observation that some subjects first attempted to find a relation between the first pair of numbers and then transferred this relation to the second pair. The relation was often a ratio simplified to 1 : x or x : 1; for the foremen, this is the identification of the scale used. Once the ratio was identified, transference to the second pair could be obtained through multiplication or rated addition (that is, addition that took into account what the initial ratio was). No distinction between multiplicative and additive solutions was drawn here because both methods depend in the first place upon a conception that is multiplicative in nature – namely, identification of the ratio. This strategy is exemplified in the protocol below, which shows the same subject transferring the ratio through multiplication in one problem and through rated addition in the other.

> J. M., illiterate, 12 years of practice in the trade, when solving the last problem above said: "Nine centimeters, three meters. Right. This is easy."
> E.: Why?
> J. M.: Because it is just, you just take three centimeters for each meter.
> E.: How did you come up with this so quickly?
> J. M.: Isn't nine equal to three times three? Then, if it [the wall] is three meters, three times three, nine. The other one here [on paper] is fifteen, the wall will be five meters. Because three times five, fifteen, this is an easy one. [J. M. found the rela-

tion – take 3 cm for each meter – and transferred it to the second pair by multiplication.]

In the 1 : 40 scale, however, which had been presented just before, he identified the relation and transferred it to the second pair (known value: 8 cm) by rated addition:

> J. M.: On paper it is five centimeters. The wall is to be two meters [values obtained from the blueprint]. Now, one thing I have to explain to you. This is not the scale that we usually work with.
>
> E.. That's right.
>
> J. M.: This one we'll have to divide. We will take five centimeters here, and here is two meters. I think this one you did on purpose. [Smiles]. I don't think they would have drawn it like this. [Other general observations.] This one is hard. One meter is worth two and a half centimeters. [Here the relation between the first pair of numbers was simplified.] Two meters, five centimeters [marking off the centimeters on the measuring stick and counting the corresponding meters]. Three meters, seven and a half centimeters. [Pause] Now, twenty-five divided into five. Three meters, seven and a half, three meters, but there's five millimeters more.
>
> E.: Uhum, five more.
>
> J. M.: If I have, two and a half of this one [centimeters] worth one meter, this one [showing 2.5 cm] divided by five. Then it is twenty centimeters [in real-life size], then you add, it is three meters and twenty.
>
> E.: What was that that you did?
>
> J. M.: This [shows 2.5 cm] is this [shows 5 mm] five times. What is it that [taken] five times gives one meter? It is all done by reasoning.

This strategy is analogous to finding the unit price when the price for *n* units is given. If one knows, for example, that 2 bags of peanuts cost 5 dollars, one can divide 5 by 2 in order to find out the price of one bag. Similarly, if 5 cm on the blueprint drawing corresponds to 2 m of real-life wall, 2.5 cm corresponds to 1 m. Thus, finding the relation is a strategy that can in principle be available both to foremen and to students.

The problem-solving strategy was identified for all subjects only when the first data pair for each scale was presented. Although three data pairs were given for scales 1 : 100 and 1 : 50, all subjects

Table 5.2.*Percentage (and number) of subjects per type of strategy for each scale*

| | Scale | | |
Strategy	1 : 50	1 : 40[a]	1 : 33.3
Foremen			
Rule of three	–	–	–
Hypothesis testing	47 (8)	25 (3)	35 (6)
Finding the relation	47 (8)	67 (8)	59 (10)
Incorrect additive solutions	–	–	–
Others	6 (1)	8 (1)	6 (1)
Students			
Rule of three	–	6 (1)	6 (1)
Hypothesis testing	–	–	–
Finding the relation	88 (14)	82 (13)	83 (13)
Incorrect additive solutions	6 (1)	6 (1)	6 (1)
Others	6 (1)	6 (1)	6 (1)

[a]For this scale, there are 5 missing observations because it was introduced into the study after 5 foremen had been tested.

worked out the scale from the first data pair and then simply transferred it to the succeeding questions on the same drawing. This way of proceeding shows that all subjects conceived of the scale as invariant in this problem situation, thus displaying a basic understanding in the problem.

Table 5.2 shows the distribution of subjects according to strategies and scales. The scale 1 : 100 was not included because all subjects simply attempted to read off the values, making the appropriate correspondence between centimeters and meters.

Some findings will be highlighted in this discussion of strategies. First, the rule of three was used by only one student when solving the problems with scales 1 : 40 and 1 : 33.3, even though all students had received instruction in this algorithm. This result replicates findings from other studies (Hart, 1981; Vergnaud, 1979) showing that the systematic cultural transmission of this algorithm in mathematics classes does not have a direct impact on students' problem-solving routines in other situations. Despite the general applicability of the rule of three, it is not very often used by students. Two (not incompatible) explanations may underlie this fact. First, the problem may not have looked like a school problem requiring school methods for solution. Scale drawing is not a topic included in the students' textbook in the chapter about ratio and proportions. Second, it is likely that the lack of use of the rule of three is a consequence of its poor ties with problem situations. If the students cannot understand that this general formula models a certain type of situation, they will not resort to it when solving problems.

Second, only one instance of the wrong additive strategy was observed in scale 1 : 40 and one in scale 1 : 33.3. This very low incidence of wrong additive solutions contrasts with the high incidence observed in other studies with adolescents. In a very similar type of problem, Karplus, Karplus, Formisano, and Paulsen (1977) observed percentages of wrong additive responses varying between 0 and 41 in seven countries. Karplus et al.'s task consisted of finding the height of a stick figure named Mr. Tall in paperclip units when the height of Mr. Short is known in paperclips, and both heights are known in buttons. Karplus et al.'s task is like a scale problem, with 1 button corresponding to 1.5 paperclips. However, the task is not presented as a scale problem, and different initial representations of the problem situation may emerge. Carraher, Carraher, and Schliemann (1986) observed in response to Karplus et al.'s task 62% of wrong additive answers in a Brazilian sample of seventh graders drawn from lower- and middle-class schools. This result is radically different from what we observed in the present study among students at the same grade level sampled only from a middle-class school. Analysis of the justifications observed in the Karplus et al. problem shows that additive responses were justified by the youngsters on the basis of a constant difference in size between Mr. Tall and Mr. Short, such as "If the difference between them is two when you measured before, it is obvious that the difference cannot change, and the answer is eight clips." In contrast, all of the students who treated the problem as a scale problem displayed correct, proportional reasoning, giving justifications like "One button equals one and a half clips. Mr. Short is 6 clips and Mr. Tall is 9 clips." Karplus et al., in agreement with the Piagetian position, treated incorrect additive and proportional reasoning as different and successive levels of development. If that were the case, given the age level of the students in the present study, we should have observed some incorrect additive responses.

An alternative explanation is possible. These two types of answer may not be ordered developmentally but simply represent the application of different mathematical models. If a situation is new, as is the one in the problem about Mr. Tall and Mr. Short, people will need to think it through in order to find out what the invariants in the situation are. Those students who conceived as the invariant the difference between Mr. Tall and Mr. Short used the corresponding additive strategy in solving the problem, whereas those who conceived of the relationship between paperclips and buttons as the invariant maintained this ratio constant. In contrast, in the present study, where it was clear from the start that the invariant was the relationship between the size of the drawing and the real-life size of the walls, incorrect additive strategies were virtually absent.

Third, finding the relation and hypothesis testing were the most common strategies among foremen. These two strategies have one thing in common: They remain closely related to the meaning of the situation. Hypothesis testing starts from known scales and tries to verify whether the numbers are in agreement with predictions from that particular scale. Finding the relation is an attempt to figure out how many centimeters on paper correspond to a meter in reality, or vice versa. These preferred strategies among foremen are therefore coherent with the picture of street mathematics we have been assembling so far: Solutions are generated through computations that allow subjects to work close to the situational relations. The two strategies are, however, rather different in their generalizability: Hypothesis testing is entirely dependent on knowledge of particular scales, but finding the relation is a general strategy applicable to any new scale.

About one third of the foremen seemed to have developed schemas for the situation so closely connected with its particulars that they did not think about other mathematical relations; the remaining two thirds understood the mathematical relations well enough to think of scales in general. It seemed reasonable to test whether this more general type of knowledge was developed by the foremen with higher levels of schooling, who could have placed their everyday experiences in a broader framework of mathematical knowledge. Subjects were grouped for this analysis according to three levels of schooling: (*a*) no schooling (all 4 subjects in this group were illiterate); (*b*) three or four years of schooling; and (*c*) five or more years of schooling. These lev-

Table 5.3.*Percentage of correct (C) and incorrect (I) answers given by foremen with different educational backgrounds*

	Scale			
	1 : 40[a]		1 : 33.3	
Schooling	C	I	C	I
None ($n = 4$)	0	0	75	25
3 or 4 years ($n = 8$)	66	34	50	50
5 years or more ($n = 5$)	0	100	20	80

[a]5 missing observations.

els were chosen because the four arithmetic operations are expected to be mastered by the end of fourth grade, and more complex mathematical models (such as algebra and proportions) are taught from Grade 5 on. The percentages of correct solutions to the problems with unfamiliar scales were then obtained for each group of subjects. Table 5.3 shows the results of this analysis. The hypothesis that schooling might have been the factor that distinguished hypothesis testers from relation seekers was not supported by analysis of the distribution of correct and incorrect responses among foremen.

A second possible explanation would be the number of years of practice in the trade. Since all foremen interviewed had at least five years of practice in the trade and the total of their years on construction sites could not be ascertained precisely, no analysis of the impact of years of practice on problem solving was attempted. There is clear evidence that it is possible for foremen to develop general representations of work situations that allow them to solve inverse problems with unfamiliar scales, but we do not know why only two thirds of them did so.

Finally, students, who did not have a pool of hypotheses to draw from and test against the data, could not work by testing hypotheses. Finding no hypothesis testers among students is therefore no surprise. What is surprising is that students' strategies showed little impact from the algorithm learned in school, there being a clear preference among students for the strategy of finding a relation – a strategy that they may have used here by analogy to the everyday situation of finding the unit price of items. In this respect, students fared better (although not significantly so) than foremen, showing a higher incidence of this general and appropriate strategy.

Despite their sound approach to the problem, students made more errors in some of the problems, a finding best understood when we look at the error analysis.

Error analysis. An analysis of the types of error observed showed the following trends. Both foremen and students had more difficulty when calculating with decimals (involved in solving problems with scales 1 : 40 and 1 : 50) than with whole numbers. Their difficulties were, however, different in nature. When foremen faced difficulty with decimals, they rounded off the results to the appropriate whole number and carried on with the new value. This way of solving the problem revealed their understanding of the situation, although we still scored their final answer as wrong. They seemed to understand decimals but had trouble in working with units smaller than millimeters, a division they understood on the ruler but may not have known the name for (some just referred to millimeters as the "little marks").

In contrast, students' difficulties were more serious, revealing themselves in two ways. First, students came up with answers that were mathematically improper because they used two decimal points (such as "three point five and a half" or "three point seven point five"), a type of error not observed among foremen. These responses not only reflect difficulty in calculating but also poor understanding of what decimal points mean. Further, students' attempts to round off results also reflected poor understanding of decimals. For example, two students rounded off the value "three point five and a half" to "three point six" although the "half" they had calculated was 0.25 cm (which added to 3.5 gives 3.75). These results are analogous to the differences between students and farmers described in the preceding chapter.

Students' greatest difficulty, however, was in attaching meaning to the result of their calculations, especially when decimals were involved – a difficulty not displayed by the foremen. For example, after computing according to a proportional model the real-life size of the wall represented by 7.5 cm in scale 1 : 50, one student interpreted the result as "three meters and 750 centimetres." Students' difficulties with decimals and the meaning of measurements were especially clear with the 1 : 100 scale, in which only adjustments of the decimal point are required. Whereas foremen displayed a perfect performance in this scale, students reading off the values showed 25% wrong answers. The students' errors indicated clear difficulties in attaching meaning to the reading. For example, a wall represented by a length of 3.3 cm was read off by some students as measuring 330 mm in real life (which is in fact the same value) and by others as measuring 33 cm. These are clearly unreasonable sizes for the width or length of a room in an apartment.

V Conclusions

This study both confirms and extends the findings described in chapter 4. Once more we observed that mathematical knowledge bears the marks of the situation in which it was acquired. Although students used a sound strategy of general applicability, they had difficulty in attaching meaning to the results of their computations. In contrast, foremen showed no such difficulty being clearly at ease with the interpretation of numbers, even when decimals were involved.

The novelty of the results observed in this study concerns the reversibility of mathematical knowledge acquired in everyday situations. Despite connections between mathematical model and situation, street mathematics is not restricted to the specific calculations carried out in everyday life. The ability to solve inverse problems with unfamiliar scales clearly demonstrates that street mathematics can go beyond everyday practice, although this may not be true for everyone. Some foremen seemed to develop very specific knowledge that reached only up to their everyday needs – like the hypothesis testers in this study, who conceived of only a finite number of scales – but the majority of foremen used problem-solving strategies that could deal with an infinite number of scales and went beyond the reach of their everyday activities.

The study of foremen's knowledge of proportions concentrated on solving problems about a content – scale drawing – that was already part of problem-solving situations in their everyday life. The novelty resided in the direction of the calculations and the unfamiliar scales. We tested for the *reversibility* of their knowledge. However, calculating with a new content, one that does not normally require computations, was not examined. The use of old abilities with new contents is traditionally called *transfer*. Can street mathematics show transfer? Isn't transfer possible only when subjects can dissociate form from content? Is street mathematics hopelessly intertwined with the situational meaning? This is the question dealt with in the next chapter.

6   Reversibility and transfer in the schema of proportionality

The picture of street mathematics developed so far shows it to be based on a *semantic* rather than a *syntactic* approach to problem solving (see Resnick, 1982, for this distinction in arithmetic). To use a semantic approach means to generate a mathematical model on the basis of relationships in the problem situation. This model of the relations in the situation is then used to guide computations during problem solving. Subjects remain aware of the meanings involved throughout their activities. In contrast, school mathematics represents a syntactic approach, according to which a set of rules for operating on numbers is applied during problem solving. Meaning is set aside for the sake of generality. In the preceding chapter, however, we saw that it is possible to achieve generality and still keep meaning in sight. Foremen using the strategy we called finding a relation demonstrated a reversible and general knowledge of scales, their problem-solving ability not being restricted to the scales they used at work.

The study of foremen's knowledge of proportional relations was restricted to one content, namely, knowledge of scales. There was no attempt in that study to look at how foremen would solve problems in

We thank the British Psychological Society for permission to reprint substantial portions of the paper "A situated schema of proportionality," which appeared in the *British Journal of Developmental Psychology, 8,* (1990), 259–268. We also thank many friends who contributed to the development of the research described in this chapter. Edvirges Ruiz and Mércia Santos introduced us in the community of fishermen. Jean Lave visited the site with us and asked significant questions. Simone Lima, Suely Magno da Silva, Joelma Alves Silva, and Telma Rocha helped us with data collection. Many fishermen contributed to this work and gave it a special flavor – particularly Jiba, who was a guide to the theory of fishing and prepared seafood so well. We sincerely thank these friends and co-workers, and also the institutions that supported this research and made the fieldwork possible: CNPq and MEC/CAPES/PADCT/SPEC.

other content areas but only to examine the reversibility of their knowledge; we asked them to solve problems that were the inverse of their daily practice. In this chapter we look at another set of data about understanding proportions. Here the subjects learning proportions out of school are fishermen, whose problem-solving ability will be contrasted to that displayed by students in the same community. The studies were designed to focus on reversibility and transfer of the schema of proportionality. The chapter is divided into four parts. In the first section, fishermen's activities and their knowledge of proportionality in two contexts are described. In the second section, we describe three studies bearing on reversibility and transfer of their knowledge of proportions. In the third section, the performances of fishermen and students on proportions problems are contrasted. In the last section, overall conclusions from these studies and their educational implications will be discussed.

I   Proportionality in fishermen's everyday life

1. Catching, processing, and selling fish

The fishermen studied by Schliemann & Nunes (1990) lived in the small community of Itapissuma, 60 km from the main town of Recife in northeastern Brazil. Their everyday activities in catching, storing, and selling fish, shrimp, and other types of seafood, fresh or processed, involve weighing, calculating prices as a function of weight, and knowing the approximate ratio of unprocessed to processed product. For instance, fishermen sell fresh whitebait to middlemen, who salt the fish and let them dry in the sun to be sold later in the market. To evaluate the price obtained from the middlemen, fishermen must have a notion of how much the fresh whitebait will "break," as they say, in this process − that is, what reduction will be observed in the final weight of the product. In other words, they must have a notion of how much fresh whitebait will in the end produce one kilo of salted, dried whitebait. Similarly, they know roughly the number of unshelled crabs needed to obtain one plate of crab fillet or the amount of unshelled shrimp needed to obtain one kilo of shelled shrimp. This is what we mean by knowledge of the ratio of unprocessed to processed seafood. This knowledge tends to be expressed in terms of how much unprocessed product is necessary to obtain *one unit of processed* seafood. This knowledge is not involved in calculations in daily practice. What matters to the fishermen is how much they are getting per kilo of fresh fish and how much the middleman is getting per kilo of processed fish in the market so that they can discuss with the middleman the price they are getting for the fresh fish. Thus, they know that 15 crabs will yield one plate of crab fillet and use this knowledge to compare the price they are getting for their crabs with the price of a plate of crab fillet in the market. But they do not calculate how many crabs they have to catch in order to get x plates of fillet or how many plates of fillet x crabs will yield.

In short, fishermen catch the fish and sell them to the middlemen. The fish are weighed as the boats come in, the price per kilo is negotiated, and the fishermen have to bear in mind the price of the processed kilo in the market. After the price has been negotiated, fishermen have to calculate the total value of their sale, given the mutually agreed price per kilo. Because of inflation, unit prices change often. Quantities fished also vary from day to day. Fishermen cannot simply rely on memory of past transactions but have to compute the total price on each occasion.

Fishermen's knowledge of pricing and processing fish offers the possibility of setting up an interesting "natural experiment" about the reversibility and transfer of mathematical models learned in everyday life. First, we know that they can calculate the price of n units, given the price per unit. They may either simply know a routine for carrying out these calculations or understand the isomorphism of measures schema in the context of the weight–price relation. In the first case, they would have a unidirectional model of the situation tied to their everyday practice. In the latter case, the model would be flexible and reversible. Second, we know that they discuss the relationship between unprocessed and processed seafood but do not use it in calculations. Can they use the computation strategies developed for prices in this new content area? The interest in this question lies in the fact that the relationship unprocessed to processed seafood is a familiar one, but there is no previous practice in calculation. In fact, when fishermen were asked to calculate using this ratio in interviews during the ethnographic phase of this study, they invariably responded by saying that this relationship was only an estimate, an average, and that the exact yield varied slightly. For example, fewer larger crabs were needed for one plate of crab fillet. Calculation was only carried out in the later interviews because they referred to hypothetical situations in which they were given information about the ratio of unprocessed to processed seafood in other parts of the country.

2. Seafood and other foods

A very large proportion of the whitebait caught at Itapissuma is not sold locally but is brought into the countryside by a second tier of middlemen who buy the salted, dried fish from the first middlemen. The second tier of middlemen bring in from the countryside other products, such as beans, dried beef, and ground cassava, that are sold in the Itapissuma market. Ground cassava is part of the staple diet, being used in local dishes with crab, barnacles, and fish broth, and is thus a very familiar product, but the processing of ground cassava is not carried out in family units in town. When cassava is ground, it is squeezed (the liquid being used for other purposes) and dried and is sold as a type of flour. As in the processing of fish, the amount of fresh cassava that goes in is proportional, not equal to the amount of ground cassava obtained. Like fishermen speaking about the ratio of unprocessed to processed seafood, subsistence farmers make further observations about the ratio of unprocessed to processed cassava, which they consider an approximation, because the yield ratios differ between the dry and rainy seasons as well as depending on whether the cassava is planted in more or less fertile soil.

People living in the town of Itapissuma buy and use ground cassava but are not likely to produce it even for their own use. They also do some gardening, but agriculture is not a main occupation carried out commercially by the fishermen. This general familiarity with the idea that food is cultivated, processed, and sold created, in the absence of more specific knowledge of yield ratios, the second context for the "natural experiment" we wanted to carry out. If one can abstract from certain details of the situation at hand, agriculture problems can be seen as involving the same type of reasoning as fishing. Calculating the price of varying amounts of cassava is not identical to but is still rather similar to calculating the price of fish. Variables here are still weight and price. However, calculating how much cassava is needed to obtain certain quantities of cassava flour is similar to knowing how many crabs yield one plate of crab fillet only if the specifics of the situation are set aside and a more general model is used as a reference.

3. Hypothetical models for fishermen's knowledge of proportions

Different hypotheses can be posed about what fishermen may learn about proportions from computing prices in their job. These hypothesis and their predictions about reversibility and transfer of knowledge are examined briefly below. In the subsequent experiments, they will be evaluated empirically.

(*a*) Fisherman may develop what has been called "procedural knowledge" (see Hatano, 1982, for an interesting discussion), that is, a type of knowledge that prescribes the specific steps in going from one point in a problem situation to another until a solution is found. They may learn to calculate the value of their catch, given the unit price, through a particular computation routine – multiplication or repeated addition – without having a more general understanding of the proportionality between the values of the two variables – amount of product and total price. If this is the case, *fishermen's knowledge of proportions should not be reversible,* because inverting the direction of the calculation requires using different steps *and should not transfer to other variables* or other content areas because the steps are specific to the situation in which they are learned.

(*b*) A second hypothesis is that fishermen may develop a conceptual understanding of the price–weight relationship. If this is the case, *they are expected to know not only the steps in the usual direction of calculation but will also be able to invert the direction of calculation. However, their knowledge of proportionality may be so entangled with the situational relations that they will not see the relevance of the proportional model to other situations.* One could say that they display conceptual knowledge of the proportionality between price and weight but do not have a general schema for isomorphism of measures situations. This knowledge would not show transfer to other contents.

(*c*) Finally, it is possible that fishermen develop a more general knowledge of proportions, understanding that certain variables (weight–price or unprocessed–processed food) stand in a proportional relationship to each other. It is possible in this case to speak of their knowledge as best represented by a more general schema for dealing with isomorphism of measures situations. *If this is the case, new contents would be handled by the schema of proportionality as long as relations were of the isomorphism of measures type.* Fishermen would then show reversible and transferable knowledge of proportions, solving equally well price–weight and yield problems in both fishing and agricultural contexts. There is, however, no reason to expect that other types of situation involving other models (product of measures or multiple proportion) would also be understood.

In order to evaluate these alternative hypotheses, we carried out three studies in which fishermen were asked to calculate values in number-of-kilos–price and in unprocessed–processed food relations dealing with both fishing and agricultural contents.

II Reversibility and transfer of fishermen's knowledge of proportions

1. Inversion and transfer of proportional relations in isomorphism of measures situations

In this study we asked 22 fishermen (21 men and 1 woman) to solve a series of four types of proportions problems.

Type 1 problems related to the actual prices each fisherman worked with. The interviewer asked fishermen how much they were getting per kilo for the product they sold and then asked how much 3 and then 5 kilos would cost. Thus, specific numbers in the problems varied from one fisherman to another, but their connection with the subjects' practice was kept constant.

Type 2 questions were also about the price–weight relationship but involved inverting the direction of the usual practice in calculating. For example, subjects were told that another fisherman was selling his fish at the rate of 75 cruzados (the Brazilian currency at the time) for 5 kilos; they then were asked to calculate how much that fisherman was getting per kilo.

Problems of types 3 and 4 involved the relationship between unprocessed and processed seafood and were both transfer tasks. *Type 3* problems were inverse questions regarding the unit yield of, for example, a kind of shrimp fished in the south that yielded 3 kilos of shelled shrimp for 18 kilos of shrimp caught.

Type 4 questions involved nonunit points in both variables. For example, subjects were told that the ratio of unprocessed to processed seafood was 7 to 4 and were asked how much a fisherman would have to catch in order to fill an order for 16 kilos of processed seafood. This last type of problem is typical of the exercises used in school for problems in proportions. Absence of reference to the unit makes the problem more distanced from everyday practice than the other types and requires greater flexibility in the use of strategies for calculating proportional values.

The fishermen interviewed in this study had from none to nine years of school experience. Since in Brazilian schools the proportions algorithm is taught in the seventh grade, only two subjects are assumed to have received school instruction on proportions.

a. Procedure

Subjects were interviewed individually on the beach where they worked. They were told we were interested in knowing how they solved some mathematics problems in everyday life. Paper and pencil were not provided, because they are not used by fishermen when they come from the sea, weigh their products, and carry out their transactions with middlemen. Interviews were tape-recorded, and notes were taken by an observer when needed. Subjects usually spoke out loud as they solved the problems, thus using oral mathematics. Their oral calculations and the interviewer's notes were later used for a qualitative analysis of the problem-solving procedures.

b. Results

The results will be discussed in terms of the accuracy of responses and the strategies used in calculating.

Accuracy. The percentages of correct responses by type of question are presented in Figure 6.1. Although the fishermen performed significantly better on type 1 than on type 2 problems (Sign Test, $p < .02$), the percentage of correct responses on type 2 questions was still high enough to demonstrate that many of these fishermen were able to solve problems that are the inverse of their everyday computational experiences with prices. Hence, they had not learned a simple

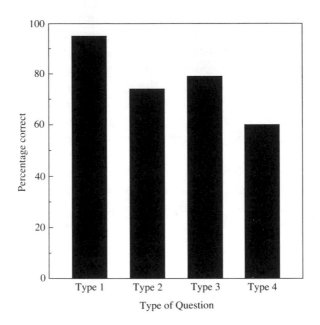

6.1. Percentage of correct responses among fishermen, by type of question

step-by-step procedure for solving proportions problems in everyday life but had developed knowledge of a more flexible nature.

For type 3 problems, 79% of the answers were correct. No significant difference was observed between the percentages of correct responses between type 2 and type 3 problems (Sign Test, $p = .50$), which were both inverse with respect to everyday experience but differed in the nature of the variables involved. Thus, if relations in the new content are known to be proportional, there is no difficulty in applying an everyday schema to problems with a new content. All errors in type 2 and type 3 problems were approximate answers that showed that subjects could not be using an incorrect additive solution in solving these problems. (For further discussion of this type of solution, see chapter 5.) For example, in the problem "5 kilos cost 75 cruzados; how much per kilo?" there were wrong answers between 12 and 19, whereas an incorrect additive strategy would generate a response like "71."

Type 4 problems were the most difficult (Sign Test comparing type 3 and type 4 questions, $p < .04$): Only 60% of the answers were correct. Increasing the distance between everyday practice and the problems presented in the experimental situation had a significant effect upon performance. The nature of this impact will be better understood when the strategies used in problem solving are analyzed in the next section.

No correlation between years of schooling and rate of correct responses was observed for any of the problem types. It must be recalled, however, that only two subjects had received instruction on the proportions algorithm at school, and thus no specific effect of instruction was expected. One of these subjects tried to use the proportions algorithm when solving a type 4 problem by writing in the sand. Because the proportions algorithm is part of a schooled, written mathematics tradition, the interviewing conditions may have made the use of the proportions algorithm less likely.

Strategies in problem solving. Noelting (1980) and Vergnaud (1983) have suggested that it is possible to identify subjects' strategies in solving proportions problems by examining the intermediary calculations they carry out. They classified solutions as three types, which are briefly described in Table 6.1. As can be seen from the description of these categories, their identification is only possible when all points in the variables in the problem are nonunit points.

Qualitative analysis of the transcribed protocols showed that subjects typically used a scalar solution when solving type 4 problems. The relationship between the scalar solution and the representation of a situation as in the isomorphism of measures model is clear. Subjects maintain the proportionality between the values by carrying out the parallel transformations on the variables and not by computing across variables.

Scalar solutions were easily found for two of the three type 4 problems, because the value asked for in the problem was a multiple of the value given, as in the sample problem presented in Table 6.1. However, one of the type 4 problems had values that made this type of solution more difficult: The desired number of kilos of processed food was neither a multiple nor a divider of the number given in the problem. Subjects struggled with the problem in their attempt to use a scalar solution. The question was how much to fish for a customer who wanted 2 kilos of processed seafood when 18 kilos of unprocessed seafood yielded 3 kilos of processed food. A functional solution would be simple indeed, but the clear preference for scalar solutions

Table 6.1.

Sample type 4 problem: In the south there is a shrimp that yields 3 kilos of shelled shrimp for each 15 kilos you catch. Now, if a customer wanted 9 kilos of shelled shrimp and ordered that from you, how much would you have to catch?

Scalar solution
A scalar solution consists of carrying out parallel transformations on the variables, thus keeping their ratio constant. An example of this solution was observed when J. A. S., 8 years of schooling, solved the problem: *You multiply three times fifteen. Each fifteen you get three kilos, isn't it? If you have three kilos, then fifteen plus fifteen plus fifteen is forty-five, and then you have three plus three plus three is nine, isn't it? Then it is nine and forty-five.*

Functional solution
A functional solution consists of finding the unprocessed–processed ratio and using it to calculate the desired value. No functional solutions were observed in this problem. The intermediary calculations would be: If 15 kilos yield 3, then one needs 5 kilos to get 1 kilo of shelled shrimp; in order to fill a request for 9 kilos of shelled shrimp, the fisherman has to catch 9×5 kilos of shrimp, which is 45.

Rule of three solution
The rule of three was discussed in chapter 5. It is a written arithmetic procedure that involves setting up the number in this problem as $3/15 = 9/x$. This type of solution was not observed in this problem.

---

displayed by all subjects made this the most difficult problem. Part of one subject's efforts to use a scalar solution is transcribed below.

> [After having attempted the problem once and failing, J. A. S. returned to it and gave a correct answer. He was asked to explain how he got it right this time.]
> J.A.S. . . . one and a half kilos [processed] would be nine [unprocessed], it has to be nine, because half of eighteen is nine and half of three is one and a half. And a half-kilo [processed] is three kilos [unprocessed]. Then it'd be nine plus three is twelve [unprocessed]; the twelve kilos would give 2 kilos [processed].

This scalar solution is both terribly clever and awkward. It is clever because it allowed the subject to find a rigorous route to solution (without estimating) through successive divisions and additions, but it is also clearly awkward in comparison with a functional solution. Table 6.2. shows a schematic representation of the calculations.

Table 6.2.

Procedure	Unprocessed	Processed
Given	18	3
Divide by 2		
Result (*a*)	9	1½
Divide by 3		
Result (*b*)	3	½
Add results (*a*) and (*b*)		
Solution	12	2

Problem: There is a kind of shrimp in the south that yields 3 kilos of shelled shrimp for every 18 kilos you catch. If a customer wanted the fisherman to get him 2 kilos of shelled shrimp, how much would he have to fish?

It is this contradiction between clear ability with numbers and apparent difficulty with a simple problem that strengthens the notion that subjects are using the isomorphism of measures model to solve problems in this novel situation – a model that they have already used in calculating in the price–weight relationship. The isomorphism of measures model is compatible with scalar solutions because values of each variables are kept separate during calculation. In contrast, a functional solution requires subjects in a certain sense to set aside reference to the variables in the problem in order to divide kilos of fresh shrimp by kilos of processed shrimp. Through this computation they obtain a number that refers to neither of the original variables but rather to a measure of the relationship between them.

c. Discussion

The data on the accuracy of fishermen's responses to the different types of problems included in this study can be taken as an indication that fishermen do in fact develop a general understanding of isomorphism of measures. They clearly demonstrated ability to invert the direction of their calculations and to transfer solutions to a new set of problems. Further, the analysis of strategies shows their strong preference for a calculating strategy closely connected to isomorphism of measures situations: As values in one variable increase/

decrease, values in the other variable increase/decrease by the same factor (and not by the same amount, as observed in incorrect additive solutions). However, we had not systematically varied the values used in the problems in order to see what impact the choice of numbers could have on performance (given that certain choices would make functional solutions easier and others would make scalar solutions easier). Hence, a second experiment was designed in which this effect was systematically investigated

2. Preservation of meaning in out-of-school strategies to solve proportions problems

The choice of a scalar or functional solution may of course be affected by which numbers are presented in the problem, as suspected by Vergnaud (1983). The use of one or the other type of solution indifferently is to be expected only if subjects do not have a preferred (or predominant) way of conceiving of proportional relations. But if fishermen conceive of proportions basically through a schema of isomorphism of measures, which generates scalar solutions, will they readily change to a functional solution when numbers are changed around? In other words, if there is a psychological representation of the situation from which the mathematical relations are understood, will subjects abandon this representation and use a solution that sets meaning aside? Will the rate of error increase in problems that are less easily solved through the scalar solution? Will fishermen use the functional solution in these problems, or will they find complex routes and stick to the scalar solutions?

In order to answer these questions, we investigated problem difficulty as a function of the relative ease of functional versus scalar solutions in a second set of proportions problems. Problem content was always the same – hypothetical ratios of unprocessed to processed seafood – and subjects were asked how much to fish for a customer who desired particular amounts of processed seafood.

a. Method

Subjects were 16 fishermen whose level of school instruction varied from none to eight years of school attendance. They were interviewed individually on the beach, as in the preceding study. They were already familiar with the interviewers, who by then had been working at the site for more than two years, and were simply asked to solve problems as other fishermen had been doing in order to show us how they calculate in their everyday lives. Interviews were tape-recorded and transcribed for qualitative analysis.

Six problems were presented to each fisherman, three that would be most easily solved through scalar solutions (termed "scalar problems" for brevity) and three that would be more amenable to a functional solution (termed here "functional problems"). For each pair of problems, one scalar and one functional, the same set of numbers was used.

b. Results

Accuracy. The percentage of correct responses on scalar problems was 83 and on functional problems was 70. This difference was not significant (t test for correlated means, $t_{df\,=\,15} = 1.518$; $.10 > p > .05$). No correlation was observed between level of schooling and performance either in the total set of problems or in the subset of functional problems.

Strategies used in problem solving. Qualitative analysis of the protocols revealed that scalar solutions were clearly preferred in both sets of problems: They accounted for 83.3% of the responses in the scalar problems and 72.3% in the functional problems. It is the fishermen's ability to use scalar solutions in functional problems that appears to explain their success in functional problems despite their greater difficulty.

Functional solutions were observed only once in the scalar problems and in 12.7% of the functional problems. The most interesting qualitative result lies in the ease with which scalar problems were solved, in contrast to the clever and awkward solutions given to functional problems. Two responses by the same fisherman, S., 49 years old with one year of school instruction, are transcribed below to illustrate this contrast.

> Scalar problem: There is a type of oyster in the south that yields 3 kilos of shelled oyster for every 10 kilos you catch; how many kilos would you have to catch for a customer who wants 12 kilos of shelled oyster?

S.: (Responds immediately) To get 12 [processed] is 40 kilos.

E.: How did you solve this one?

S.: It's because 3 times 4 is 12 [4 is the scalar factor; the constant in the function is 33.3]. The 12 [kilos] will be 40.

Functional problem: There is a type of oyster in the south that yields 3 kilos of shelled oyster for every 12 kilos you catch; how many kilos would you have to catch for a customer who wants 10 kilos of shelled oyster?

S.: Twelve kilos to give . . . ?

E.: Three kilos of shelled oyster. How much do you need to get 10 kilos of shelled oyster?

S.: . . . On the average, 40.

E.: How did you solve this one?

S.: It's because we make it simpler than using pencil.

E.: But how did you figure out that it's 40 for 10 kilos shelled?

S.: It's because 12 kilos give 3; 36 give 9. Then I add 4 to give 1. [Note that the constant in the function is 4; the subject knows that 3 × 4 is 12, but he does not compute the solution by using this knowledge directly; scalar transformations are calculated until he finds the unit value which he uses to complete the desired 10 kilos.][1]

The results of this study are simple indeed. Fishermen continued using the isomorphism of measures model of situations even when numbers were chosen so as to make functional solutions easier. Since functional solutions do not relate to the situational meaning as scalar solutions do, they were not generated by the fishermen even if they looked easier from our viewpoint. This finding goes against the expectation that it is easy to change the nature of the solution to proportions problems by manipulating numbers. More important, they attest to the generative and semantic nature of fishermen's knowledge of proportions and confront the idea that everyday knowledge of mathematics is procedural, being restricted to the repetition of steps learned to solve the specific problems that appear in the work setting.

---

[1] The sequence of values used leads to the solution without recourse to the functional computation. A sequence of scalar transformations (multiplying by 3, dividing by 9, and adding these two results) is sufficient.

3. A contrast between solving problems about fishing and about agriculture

These two initial studies demonstrated that fishermen were able to solve inverse and transfer problems involving the schema of isomorphism of measures. But all the problems were about relationships between variables that were of importance to the fishermen in their everyday work. A certain level of generality is demonstrated by their ability to solve the awkward functional problems in the transfer situation, but it is conceivable that their understanding will still be restricted to the life situations they are so often concerned with. For this reason the generality of their understanding was further tested when we presented fishermen with a parallel set of problems about variables from the agricultural context.

a. Method

Four new sets of questions were devised for this study, in which all of the questions were about agriculture. In the first set of questions, three problems *inverting the weight–price relation* were asked. For example, fishermen were told that a farmer was selling 25 kilos of beans for 75 cruzados and were asked to figure out how much the farmer was getting per kilo. In the second set of problems, three *questions were posed about the ratio of unprocessed to processed cassava, all of them in the usual direction of calculation, from smaller to larger.* For example, fishermen were told that a farmer was using 7 bags (a bag is a unit of weight corresponding to a fixed number of kilos, which may vary locally) of cassava to obtain 1 bag of ground cassava; the question would be posed about how much cassava would be needed if he had to produce 5 bags of ground cassava. In the third set of problems, fishermen were still asked to solve *problems about the ratio of cassava to ground cassava, but the direction of the questions was the inverse of everyday experience with calculation, going from larger to smaller.* For example, one of the three problems in this set asked how much cassava was needed to produce 1 bag of ground cassava if the farmer was obtaining 3 bags of ground cassava when he used 18 bags of the unprocessed root. In the fourth set of problems, *questions referred to nonunit points about the ratio of fresh to ground cassava.* Four problems were included here, two most easily solved through a scalar approach and two more easily solved through a functional approach when both strategies are available to the subject.

Three of these four sets of problems parallel those used in Study 1. There is no parallel to the problems using familiar prices included in Study 1, because the products in this study were agricultural. Further, this third study included a problem type not used in Study 1: direct problems about the ratio of unprocessed to processed food. Although this type of question was not important in the first study (in which relationships were more familiar), it seemed quite important in the agricultural context, in which subjects might not think of the relationship between the two variables as a ratio but could be using an alternative conception. Direct problems render the identification of such an alternative conception easier.

The specific numbers used in the problems in this study were the same numbers used in the preceding experiments. By keeping the numbers constant, we could approach the same subjects and later carry out comparisons between their solutions to problems in the two contexts. Despite the fixed order of testing and the increasing familiarity between experimenters and fishermen, these comparisons can still show whether or not performance drops sharply when a new problem content is introduced.

We were able to interview 19 subjects who had already been interviewed in one or the other of the first two studies. As before, they were interviewed on the beach without being offered paper and pencil. Interviews were tape-recorded and later transcribed for analysis. Various intervals had occurred since the preceding interviews, but none were longer than a month.

b. Results.

Since the main purpose of presenting these new sets of problems to the fishermen was to verify whether they would still solve proportion problems successfully in a new content area, this analysis concentrates on the accuracy of solutions. The percentage of correct responses per type of problem in the agricultural context is presented in Table 6.3. It is clearly seen in this table that fishermen were still quite accurate at solving proportions problems involving variables in agriculture.

Table 6.3. *Percentage of correct responses given by fishermen to problems about agriculture*

Problem type	% correct responses
Inverse weight–price	84
Direct ratio unprocessed–processed food	91
Inverse ratio unprocessed–processed food	83
Ratio unprocessed–processed food, nonunit values	75

Figure 6.2 presents comparisons of the same types of problems in the fishing and in the agricultural context. The great similarity in performance across contents is quite apparent in the figure. It can also be noticed in the figure that the direction of the differences across problem types did not remain constant. Higher proportions of correct responses were observed in the agricultural content problems in two problem types, whereas in the third type more correct responses appeared in the problems with fishing content. This instability of direction is suggestive of random, not systematic, variation as a function of problem content – an interpretation that was confirmed statistically when no significant differences within problem types and across contents were observed even for the most difficult type of problem, according to *t* tests for correlated means.

It could be hypothesized that this lack of statistically significant difference resulted from a ceiling effect, because performance was clearly high for all problem types. This hypothesis is not supported by further statistical analysis, which investigated the effect of problem type within the problems with agricultural content. In this analysis, two types of significant differences were obtained. First, problems not involving the unit were significantly more difficult than the easiest group (but not the other two groups) of problems. For this analysis, the scores for each group of problems were adjusted to the same scale (with maximum scores equal to 4), and problem type was analyzed through an ANOVA with repeated measures. A significant overall effect of problem type was observed. Comparisons between the means showed a significant difference only between problems not involving unit points, which were the most difficult, and the direct problems on

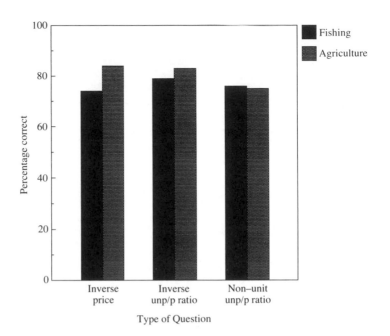

6.2. Percentage of correct responses among fishermen, by type of question and problem content

the ratio of unprocessed to processed cassava, which were the easiest. Second, within this more difficult group of problems, a significant difference was observed between problems that allowed for a simple scalar solution and those that would be most amenable to a functional solution. The mean number of correct responses (out of two) for the scalar problems was 1.84 and for the functional problems was 1.37, the difference being significant according to a t test for correlated samples ($t = 2.96$, $df = 18$; $p = .008$).

In summary, solving proportions problems about agricultural variables was not significantly more difficult for fishermen than solving the same types of problems about fishing. Their degree of success was quite high, varying between 75% and 91% correct responses in the different groups of problems. We may thus conclude that fishermen, although using a semantically based model to derive solutions to proportions problems, do not display knowledge that is so content-bound that no transfer is possible. They clearly showed their ability to transfer their model of the weight–price relation to other variables in the fishing context and to similar variables in the new problem context of agriculture. This lack of difference across contents was not the result of a ceiling effect. Although no significant differences were observed across contents, some significant differences appeared within the agricultural content, direct problems being significantly easier than the most difficult problems (the ones involving nonunit points in the variables) and scalar problems being easier than functional problems. Thus, the same tendencies in level of difficulty for the different types of problems were observed in this new content, *bringing further confirmation to the idea that fishermen develop a schema for isomorphism of measures situations that can be used in the solution of new but similar problems.*

We turn now to a comparison between knowledge of proportions developed out of school and school knowledge of proportions.

III Strategies in solving proportions problems in and out of school

In chapters 2, 3, and 4 we saw that school arithmetic and street arithmetic differ in several ways. One of these differences was that school arithmetic is distanced from meaning and proceeds in a rule-based fashion whereas street arithmetic preserves situational meaning and uses this meaning in generating solutions. Our observations of students solving proportions problems make the picture of school mathematics a bit more complicated. When we observed students solving problems about scale drawing (chapter 5), they did not use the school-taught algorithm for solving proportions. They resorted instead to the same type of solution used by foremen, whose mathematical knowledge was mostly developed out of school. A possible explanation for this "forgetting" of a general strategy is that it is poorly connected to previous knowledge and problem situations. If the relevance of the proportions algorithm is not understood, it is not going to be used, despite its ample possibilities of application. Knowing how and knowing when to use a mathematical technique are clearly distinct aspects of knowledge (see, for example, Bryant, 1985). However, it is also possible that students did not see the relevance of the rule of three to the scale problems because the situation was much too novel for them. They had not worked with scale drawing and may never have thought about the possible connection between scales and the rule of three.

The variables investigated in this study afforded a good opportunity to look at other students solving problems with familiar contents without having had experience with the same type of problem at work.

School instruction on the rule of three is given without any concern for what type of representation of proportional situations students might already have developed. This is hardly a characteristic of instruction in the small town of Itapissuma but rather reflects the beginning stages of research on the psychology of mathematics learning and its distance from actual classroom teaching. It can reasonably be expected (although there was no independent verification on this point) that Itapissuma schoolteachers taught their seventh graders the rule of three without considering the possibility that they already had an everyday schema of proportionality based on the isomorphism of measures model. Thus, learning the proportions algorithm in school could result in learning how but not when to use it.

As discussed earlier, the rule of three is not easily connected to the isomorphism of measures model, because it requires setting meaning aside. Like functional solutions, the rule of three involves computations across variables. In contrast, preservation of meaning is obtained by parallel transformations of the variables without computation across variables. Given the existence of two sources of knowledge about proportion – activities in and out of school – with two different approaches to problem solving and no systematic attempt to establish a relationship between them, it is likely that the two practices will form separate systems of arithmetic, just as oral and written calculation do. It is also likely that everyday knowledge will compete with the school-taught procedures when pupils attempt to solve problems involving isomorphism of measures situations. It is even possible that neither the school nor the out-of-school method will be properly used – the first as a result of poor teaching/learning, and the latter as a result of its exclusion from the school situation. In order to examine the question of the relationship between everyday and school knowledge of proportions in a content area already familiar to students, a fourth study was carried out in the same fishing town with students as subjects; and their performance was contrasted to that of fishermen.

a. The search for suitable subjects

The town of Itapissuma lives off fishing and marketing fish. Only relatively recently have industries developed in town and in the surrounding areas that offer new sources of employment, so it is still difficult to find young men who do not have contact with fishing and marketing fish in Itapissuma. Youngsters start to go out in boats as early as age 12, and many young men who hold other jobs go out fishing on weekends to increase their income. To find a group of male students who had no fishing experience would have been hard indeed, and if found, the group would be atypical. For this reason, we decided to work with female students as a comparison group.

Women do not go out in boats to fish. When they fish, they work in the marsh areas catching barnacles, oysters, crabs, and other shell-fish. They are not called *pescadoras* (fishers), but *marisqueiras* (oysterers). Fishing from boats is a regular occupation in town; it is neither important nor looked down upon. Being a *marisqueira* is an occupation that indicates very low family income: The work is ex-tremely hard, the schedule varies with the tide, and housekeeping is interfered with. For this reason, it was easy to find female stu-dents who did not have firsthand experience with fishing and sell-ing fish – although their fathers and brothers most certainly did, and they were surrounded by fishing-related activities in their every-day lives.

The life-style in this small town certainly must have required our young women to go to the market many times and shop. They were likely to have carried out computations using the price–weight rela-tionship. In contrast, they were unlikely to have worked with the ratio unprocessed to processed seafood. This type of question would deal with familiar themes but unpracticed problems. This was certainly a transfer task for them, as it was for the fishermen.

A group of 22 secondary students attending the newly founded grade-school teacher education program in the town was interviewed for this study; their performance will be compared to that of the fishermen in Study 1 for inverse problems and in Study 2 for one scalar and one functional problem. In all cases, identical problems are posed. The students' level of instruction was in the range from 9 to 11 years of instruction – that is, between two and four years beyond the point where they received instruction on proportions. Their ages ranged from 14 to 21, with an average of 17.1. The fishermen's level of instruction ranged from 0 to 9 years, with an average of 3.9 years; their ages ranged from 15 to 63, with an average of 36.4.

b. Procedure

The students were interviewed individually in their school. Interviews were tape-recorded and transcribed for qualitative analysis. They were asked to solve eight problems that would give an overview of their performance in comparison to that of fishermen: (*a*) three inverse price (type 2) questions and three inverse (type 3) questions about the ratio of unprocessed to processed seafood, and (*b*) two type 4 proportions problems, one favoring a scalar solution and one favoring a functional solution. The same sets of numbers were used in the type 4 problems.

The students were instructed to solve the problems using whatever means they wanted and to talk aloud while trying to solve them. As this was a school setting, paper and pencil were available on the desk, but students were not instructed to use them. Whenever necessary, after a solution was reached they were asked to explain how they had obtained the result.

c. Results

Accuracy. The percentages of correct solutions observed in each type of problem are presented in Figure 6.3. Students gave significantly more correct answers to inverse price questions than did fishermen (χ^2 $1df = 7.38$; $p < .01$). For inverse ratios of unprocessed to processed seafood, the difference between the two groups was not significant (χ^2 $1df = 0.667$; $p < .30$).

The difference in the students' performance in inverse price questions and in inverse ratios was significant (Sign Test, $p < .05$). This result supports our assumption that students developed their proportionality schema in commercial transactions and that the latter task was a transfer task for them. However, they were still able to transfer their knowledge of proportions from price–weight relations to the task of finding the ratio of unprocessed to processed food. Their high rate of success also demonstrates that most students understood the multiplicative nature of the relationship between unprocessed and processed seafood.

Students performed similarly to fishermen on the scalar problem (χ^2 $1df = 0.14$; $p > .50$), but their performance in the functional

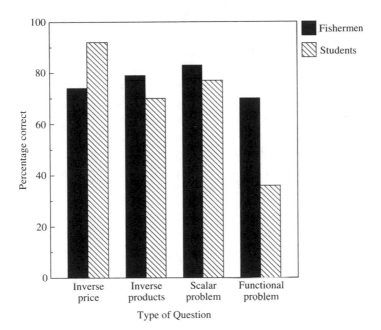

6.3. Percentage of correct responses among fishermen and students by type of question

problem was worse than that displayed by the fishermen. However, the difference between the two groups in the functional problem was only marginally significant (χ^2 1df = 2.70; .10 > p > .05). Of greater importance is the fact that whereas fishermen's performance was not significantly worse in functional problems than in scalar problems, students showed a significantly lower percentage of correct responses to the functional problems (Sign Test, p < .002).

Strategies used in problem solving. Qualitative analysis showed that among students *all correct solutions to the scalar problem were scalar solutions* and that the proportions algorithm, which they had had instruction on, was not used by any student to solve the scalar problem. In the functional problem, 62.5% of the correct responses were obtained through scalar solutions, and 37.5% through either functional solutions or the use of the proportions algorithm.

These results support the hypothesis that students continued to use their everyday knowledge when facing proportions problems despite having received instruction on the proportions algorithm. When scalar solutions were easily used, they were responsible for all of the correct responses observed. When they were harder to use than the functional solution, they still constituted the strategy used in more than half the correct solutions. It seems safe to conclude that students preferred the scalar solution because they understood it better: It is a solution that preserves meaning and allows for better monitoring while calculating. It also seems safe to conclude that when school-taught procedures come into conflict with a previously known out-of-school model for proportionality, school procedures are poorly learned and quickly forgotten.

IV   Conclusions

In summary, this series of studies demonstrates that the concept of proportionality does not have to be taught. It can develop on the basis of everyday experience. The resulting conceptual schema – based on situations of the isomorphism of measures type – models relationships in everyday situations but clearly surpasses the procedures used in everyday practice. It is not unidirectional, as everyday practices tend to be, and it can be applied to new situations.

These results clearly contradict the idea that street mathematics is the product of concrete thinking and that it generalizes poorly. Both flexibility and transfer were more clearly demonstrated for everyday practices than for the school-taught proportions algorithm. Research in both Britain (Hart, 1981) and France (Vergnaud, 1983) has indicated that students do not often draw on the proportions algorithm to solve proportions problems. But why should this be the case? Why should such a useful procedure of great applicability be so readily abandoned? This chapter has brought some answers to the question. It seems that everyday procedures, which are likely to be already available to students before they are taught the algorithm, compete with the algorithm. The conflict stems from the fact that the everyday knowledge uses calculation procedures in which variables are kept separate. No calculations across variables are carried out. The school-taught procedure violates this principle. Thus, it is not easily coordinated with students' previous knowledge. This conflict may well be at the root of poor learning of school methods for solving proportions problems.

7　Reflections on street mathematics in hindsight

In this concluding chapter, we discuss from a more theoretical perspective some of the points that were raised and documented through empirical research in the preceding chapters. The discussion is organized around two focal points: (*a*) psychological considerations about street mathematics and (*b*) lessons for education.

I Psychology and street mathematics

We started our analysis of street mathematics by relying on working definitions that took for granted what mathematical activity is. We decided to call "street mathematics" that mathematical activity that is learned and carried out outside school. This working definition allowed us to proceed to some empirical investigations. We can now think back on the observations we made and try to characterize mathematical activity more clearly. This is not a mere exercise. It is an attempt to find the boundaries of a psychological phenomenon – it is like defining affect, motivation, drug addiction, or reading ability. A psychological definition is, in fact, a local theory about the phenomenon; it is a commitment to treat some events as belonging together and to try to find generalizations that apply to all instances of that phenomenon. We start by reflecting on what mathematical activity is and see how this general notion relates to the more specific phenomenon we want to focus on: street mathematics. We then look at the psychological processes involved in mathematizing an object.

1. Defining mathematical activities

Defining any activity in general terms is not a simple matter that can be sorted out by the application of one criterion. For

127

example, an activity cannot be defined simply by its form; the gesture of casting a vote in an urn and putting a letter in a mailbox may have the same form, but the activities are certainly different. Understanding an activity requires considerations about form, attitude, goals, and relationships between the activity in question and other activities. Below we consider two aspects that have been suggested as essential criteria for defining mathematical activity and see how they apply to street mathematics.

a. Mathematical goals and their relation to particular observations

Mathematical activity is generally recognized as postulational thinking. It is not concerned with observation, experimentation, or causation but rather with deduction. In other words, mathematical activity is not carried out in order to discover relationships about empirical events but is an exploration of relationships between representations. It is a process of making inferences that starts with representations that make possible the use of formalization. Mathematical activity is not carried out in search of empirical validity and cannot be falsified by empirical facts.

How does this definition apply to street mathematics? Is it possible to conceive of interactions in a street market, for example, as involving a process of inference making that respects the rules of the formalization being used and leads to conclusions that cannot be rejected by empirical data?

We think the answer to this question is yes. People engaged in transactions in a street market behave in ways that are regulated by both social rules and logical rules. Neither type of rule is like an empirical law, which indicates certain sequences of events to be necessary. Social and logical rules acting together in the marketplace are mutually recognized obligations to behave in a consistent but not inevitable way (see Kaye, 1982, for further discussion of social rules). This means that there is a moral but not mechanical necessity of following an implicit contract that applies in the marketplace. When people calculate prices in a street market, for example, what they want is to make sure that they are paying the fair price, given the rules of the game. If one mango is marked as costing 50 cents, they can deduce that the price of five mangos is $2.50. Paying more than that is not fair; paying less is a good buy, a sale. People assume a particular connection between amount purchased (number of items, weight, etc.) and price. If I buy one fruit, the price is x; if I buy two of them, the price is (ought to be) $2x$.

Why is that so? When shopping at the supermarket or on a street corner, shoppers seem to represent the situation as "price is a linear function of quantity purchased with the y intercept equal to zero" (that is, if I buy nothing, I pay nothing). By representing reality in this way, shoppers deduce new prices that were not given in the initial agreement with the seller. According to this logic and the social rules of this situation, shoppers are allowed, for example, to confront the seller if the price charged exceeds what is fair according to the linear model. Shoppers cannot tell sellers that the unit price is wrong (they may say it is high, it is absurd, it is robbery, but not that it is wrong). In contrast, given the unit price and the assumed linear model, a shopper can tell a seller that he or she calculated the price of five mangoes wrong. If the total price calculated by the seller deviates from the linear model, it is the seller's calculation that is treated as wrong, not the linear model. The customer can use mathematics to "correct reality" if the seller tries to charge an excessive price.

This analysis clarifies the postulational nature of reasoning in street mathematics. Conclusions from the model (i.e., the new prices calculated) are tested for their validity and for their logical consistency within the assumed model, not for their agreement with empirical observations. The social agreement about the way in which quantity and price relate in the marketplace is not a necessary sequence of empirical events but involves a conjunction of social and logical rules.

Different rules may be applied in other situations. A phone bill, for example, is typically charged as a linear function of the number of units used, but the y intercept is not zero. In many countries there is a flat rate, which corresponds to rent for the apparatus and represents a constant to be included in the bill. If no calls are made and no units are used, the phone bill is not zero. A third type of model may be used, for example, to determine the value of an electricity bill, especially if the community is under pressure to save energy. In this case, companies may decide not to use a linear model but to increase the amount charged by unit used after a certain value. In all of these examples, once a model is assumed as a description of the situation, conclusions can be drawn about the price for any quantity. The social and logical rules still apply in the same way despite variations in the mathematical model assumed. This means that when entering a new contract, customers may need to test the linear model, for it is not the only possible model. Testing the linear model in a new shopping situation is to test *its fit*, not correct it, because the test does not affect the linear model but rather assumptions about the situation. This is "almost experimentation," as Lakatos (1981) would say, but it is not experimentation; it is simply a verification of the appropriateness of an old model in a new situation.

In summary, mathematical activity in the marketplace involves reasoning that is determined both socially and logically. Prices are socially constructed realities; the use of a linear model or any other is negotiated in politics, not in mathematics. The reasoning that goes on after the model is fixed is postulational, not inductive. It is the process of drawing conclusions from the model, once a model is assumed, that constitutes mathematical activity in the streets.

b. A "theoretical mathematical" attitude toward knowledge

Cole and Scribner (1974), Luria (1976), and Scribner (1975) have enriched the analysis of the nature of logicomathematical reasoning with further considerations. They have argued that *some people treat premises as statements about reality.* If asked in an experiment to form conclusions from premises, they may give correct answers. It will *look as if* they arrived at the conclusions from the premises, but they may actually have used their past knowledge about the world in order to answer the questions. If they were given counterfactual premises, they would be unable to form any conclusions. They would lack a "theoretical mathematical" attitude toward problems. These investigators independently observed that unschooled adults were able to reach correct conclusions from premises in syllogisms if the premises were in agreement with their experience. If the premises contradicted their experience or dealt with unknown facts, subjects made errors or rejected the possibility of coming to a conclusion on the basis of the premises.

Luria (1976) used this notion of a theoretical attitude toward the basic premises in a study on arithmetic problem solving. His subjects were asked to calculate the distance between two towns, A and C, given the distance between each town and an intermediary town – that is, given AB and BC. The distances given in the problem were wrong: They placed town B as halfway between A and C when B was in fact farther from A than C. According to Luria, unschooled subjects rejected the problem and refused to search for a solution. He attributed this rejection to their lack of a theoretical attitude toward the problem; the subjects considered it impossible to treat the data as premises because the data contradicted their experience. In another type of problem, Luria asked his subjects to calculate how long it would take to travel by bicycle between two towns knowing that the time a person takes to walk between the town is x hours and that a bicycle goes six times faster. In this case, subjects had to reason about a situation they had no experience with because they did not have bicycles. Unschooled subjects either (a) rejected the possibility of solving the problem by placing it outside their experience or (b) refused to take the data in the problem as premises because the time needed to walk from one town to the other went against their experience. In both cases, they failed to display a theoretical attitude.

These results would suggest that street mathematics – the only type of mathematics that Luria's subjects knew, for they were unschooled – is restricted to reality. Street mathematics would, in this case, conflict with the theoretical attitude. However, interpreting Luria's and Scribner's results is not a simple matter, as Dias (1988) was later able to show. When subjects in a psychology experiment are presented with premises that go against their experience and are simply asked to reason according to the premises, they in fact have difficulty in carrying out this request if they are young children or unschooled adults. The conflict between what they know to be true and what the experimenter asks them to consider may be solved in different fashions by different subjects: (a) A small percentage of subjects succeeds in doing what the experimenter requests them to do and considers only the premises in their responses; (b) others ignore the premises and consider only their experience in their responses; and (c) still others reject the problem. Dias has shown that there are circumstances under which these same groups of subjects may be willing to treat premises as such even if they go against their experience. If the experiment is carried out in a way that makes it clear that their experience is to be looked upon as irrelevant, young children and unschooled youngsters reason on the basis of premises and ignore their experience. In a series of ingenious experiments, Dias (1988) showed that the theoretical attitude could be produced in an experiment cither by asking subjects to engage in make-believe play (Dias & Harris, 1988) and imagine that the given premises are true on another planet or, more generally, by inducing children to create a mental world that they temporarily consider in their reasoning. Thus, the theoretical or mathematical attitude *may not be a characteristic of subjects but may or may not be produced in social interactions, depending on how these interactions are structured.*

In our experiments, we often posed problems that involved imagined situations. In some cases, we knew that the problems involved assumptions that were contrary to the subjects' experience. For example, when asking fishermen about ratios of unprocessed to processed food, we used ratios they knew from their experience to be incorrect. In these cases, we introduced the problems by asking them to imagine that there was a kind of shrimp in the south (our studies were carried out in the northeast) that had such-and-such a ratio. Fishermen often made comments about how big the shrimp must be to yield such a good ratio, but still they solved the problems. No problem rejections were observed. Given the appropriate interaction conditions, they displayed a theoretical attitude toward the unusual values, treated them as premises, applied the linear model, and reached valid conclusions.

Thus, we can see that the descriptive criteria that have been used to identify activities as mathematical are satisfied by street mathematics. However, although this analysis adds something to our view of what street mathematics may be like, it does not contribute to our understanding of the psychological processes involved. We still do not know how street mathematics is organized and how it is learned. It is these psychological aspects of street mathematics that we discuss in the following section.

2. Psychological processes in mathematizing situations

We have seen some aspects of mathematical activity that have to do with the form of mathematical thinking and its purpose. If we now think about its object, we will see that mathematics is concerned with number and space. However, it is not enough to define what the objects of mathematics are; one also needs to consider *at what level* the objects are being dealt with (Vergnaud, 1987). Gal'perin and Georgiev (1969) proposed that one should examine the passage from a "natural" to a "mathematical" object in order to understand what is essential to mathematical activities from the psychological standpoint. In the following sections, we discuss three aspects of this passage from the natural to the mathematical object, focusing on number rather than space because all the work we reported on earlier has to do with number. We discuss three steps in the process of mathematizing objects: isolating dimensions, defining units, and modeling the situation.

a. Isolating dimensions

To transform a natural into a mathematical object, one must pull out one (or more) of its dimensions and then use it (them) in classification or quantification. Representation of the natural object in mathematical form requires a choice to consider some of its aspects (the dimensions pulled out) and ignore others (those that will not be used in classification or quantification). This is essential for the use of mathematical forms – such as classes and quantities – in making inferences. It is such a basic process that we often take it for granted – even young children demonstrate their ability to isolate dimensions for particular purposes. In fact, this ability is necessary for the use of many words in natural language. For example, children could not learn the meaning of terms like "red" and "blue" if they were not able to consider color alone and ignore other aspects of the object.

However, several decades of work on conservation have shown that isolating number from space is not simple for 4- and 5-year-old children. Although the understanding of conservation does not seem to be an all-or-nothing question, as Piaget and his collaborators seem at the beginning of their studies to have expected it to be (Piaget, 1952), and although the percentages of children giving conservation responses at some age level vary considerably, depending on the experimental procedure (Light, 1986), 4–5-year-old children in all experiments show considerable difficulty in isolating number from space. In all of our studies of street mathematics, the separation between number and space is in fact taken for granted, and this level of abstraction is certainly surpassed by subjects answering questions in any of the studies we presented.

At a more advanced level, isolating dimensions corresponds to the concept of "variable" – that is, one dimension that may take different values. Because we did not concern ourselves with the concept of variable in any of these studies, we shall not discuss this point any further.

The ability to single out one dimension is only the first step in the process of understanding number. The next step, quantification, is discussed below.

b. Choosing a unit and quantifying

According to Vergnaud (1987), the concept of number is closely associated with the concept of measure – cardinals are measures. In order to quantify a dimension, it is necessary to choose a unit of measurement and to apply this unit to the object being measured. However, when children count objects in a set, *they may not be aware of the fact that objects are being taken as units*. Objects become "natural" units in counting without much need for understanding; this concept may be bypassed in counting discrete objects. It is only when a variety of units can be applied in counting that the choice of units comes into question. We observed in a previous study (T. N. Carraher, 1985) that 5–6-year-old children may be able to count but still fail to understand the concept of unit. In that study, Brazilian children were presented with two rows of play money with the same number of coins in the rows but different denominations of coins – say, one row with five one-cent and the other with five ten-cent coins. We then asked the children two questions: (*a*) How many coins in each row? and (*b*) Who will be able to buy more candy at a shop, the child with the row of one-cent coins or the child with a row of ten-cent coins? Or will they be able to buy the same amount of candy? In order to answer the first question properly, it is enough to count the coins and apply the cardinality principle. In order to answer the second question properly, the value of the units must be taken into account. Not all Brazilian preschoolers who could count accurately and use the cardinality principle could also realize that if the number of coins is the same but one row has more valuable coins, then that row has more money in it. Nunes & Bryant (1992) replicated this observation with 5-year-olds in Britain, who could count better than the Brazilian 5-year-olds but who still made plenty of inference errors about the values when units of a different size were concerned: 56%

of the responses given by 5-year-olds were wrong when coins of different values were used.

Gal'perin and Georgiev (1969) also looked at this concept, in a different type of experiment. They asked young children to put sugar in a bowl using a small spoon and then to predict how many spoonfuls of sugar were in the same bowl if they used a large spoon instead. If children understood the concept of unit, they would realize that even though the amount of sugar was kept constant, the number of spoonfuls would have to be smaller. Young children (aged 4–5 years), however, either thought that the number of spoonfuls would be the same or that it would be larger with the larger spoon.

It is thus clear that young children have difficulty with the concept of unit. However, users of street mathematics must overcome these difficulties, because this level of understanding of number is required in practices using street mathematics. In order to count money in the market, for example, the relative values of the different notes and coins must be taken into account. There is evidence that this understanding can be developed in the total absence of schooling and thus is part of the knowledge of users of street mathematics. Illiterate Brazilian adults, who are quite familiar with a money economy, *made no errors* in questions comparing buying power when the same number of notes of different denominations was used in building the total quantities (T. N. Carraher, 1985).

Taking units into account in the measurement of length may be more difficult than taking them into account in the context of monetary transactions. Saxe and Moylan (1982) showed, for example, that some, but not all, unschooled Oksapmin adults demonstrated an understanding of the concept of units in the measurement of length. Oksapmin children and adults wear on the shoulder or around the neck a string bag, which is measured by putting the hand into the bag and describing how far along the bag reaches.

The size of the bags is referred to as "knuckles," "wrist," "forearm," "inner elbow," "biceps" or "shoulder." . . . A particularly interesting feature of this system is that, unlike conventions for measurement in the West, units are not equivalent across individuals since people's arms vary in size from one individual to another. (Saxe & Moylan, 1982, p. 1243)

This means that the same reading for two bags – for example, "shoulder" – does not refer to the same length if one bag is measured by a child and the second by an adult. Saxe and Moylan devised a task in which Oksapmin school children and unschooled children and adults were asked to make inferences about the length of bags when they had been measured against units of different sizes. For example, they were asked whether a shoulder bag, having been measured by a child, would be a shoulder bag when measured by an adult. Their results showed that the question of size of units is not trivial in this context. Unschooled adults outperformed schoolchildren *with as much as six years of schooling* but still did not perform at ceiling level on this task. Only 64% of the unschooled adults showed three or four correct responses out of four questions.

In summary, although the concept of unit can be bypassed when natural objects are counted, several activities in street mathematics require an understanding of this concept. Users of street mathematics not only abstract dimensions but also use the concept of units in their activities. If they did not use this concept, they could not count money. It appears that the concept of units is more easily learned in some situations than in others. Whereas illiterate Brazilian adults had no difficulty in taking the value of units into account when making comparisons between amounts of money, Oksapmin adults showed some difficulty with this concept. However, the difficulty of coordinating several types of units in the measurement of length can be overcome in the practices of street mathematics without the support of schooling. Foremen working in the construction industry, even if illiterate, did not have trouble with conversions between meters and centimeters and moved back and forth with great ease. Thus, these two initial steps in the passage from the natural to the mathematical object are clearly accomplished in street mathematics.

We turn now to a third psychological process involved in mathematical activity: the generalization of relationships across situations, which relates to what mathematics educators call "modeling."

c. Modeling and solving problems

In order to analyze the psychological questions involved in modeling and solving problems, we want to discuss briefly what modeling is and then look at three theoretical concepts we find helpful in our attempt to understand street mathematics. These theoretical concepts converge toward a theory of cognition in practice and deal with the same basic problem: "generalization" or "transfer" across instances of practical experience. (For an excellent discussion of the literature on transfer, see Lave, 1988.)

What is essential to the idea of modeling? A simple example can be used to illustrate this aspect of mathematical activities. We know that young children can solve simple addition and subtraction problems in the presence of objects by carrying out (or even imagining) the actions mentioned in a problem and counting the resulting sets. For example, children can be asked to solve the following problem: "There were four bricks in a box and I put in two more bricks; how many bricks are there in the box now?" Young children can solve this problem by *actually carrying out the actions described – that is, by putting 4 bricks and then 2 bricks inside the box and then counting all the bricks in the box.* They can even solve this problem when they are asked about a hypothetical box, perhaps by imagining what happens when you add two more bricks to the four already in the box (Hughes, 1986). In these examples, children are counting bricks to solve problems about bricks. Suppose, however, that there were no bricks available and children were asked to solve these problems. If they were to represent the number of bricks on their fingers and count the fingers, they would be representing relationships between bricks through relationships between fingers. Using fingers to represent bricks means to recognize that results obtained with fingers are identical to results obtained with bricks. In other words, children solving problems in this way behave as if *the objects that are counted do not matter. It is the relationship between quantities that matter.* As Vergnaud (1987, p. 50), suggested, at some level the children must know that "the truth of $3 + 3 = 6$ does not depend on marbles, sweets or steps." Although simple, this is an instance of modeling, of applying techniques known in one field to obtain results in another.

Modeling is not concerned with the quantification of one object but with the mathematization of situations. The *relationships between quantities* are the essential aspects of a model. Thus, modeling is at the center of street mathematics because mathematical activities outside school are related to solving problems in everyday life. In order to understand the psychological processes involved in street mathematics, we need a theory that allows for the analysis of situations and their representations. In the following paragraphs, we look at three theoretical inputs that can be helpful in this analysis.

Three ideas and one theory of cognition in practice

A scheme of action and a schema of a situation: The Piagetian distinction between "scheme" and "schema" offers a good starting point for further reflections on street and school mathematics. These concepts have been applied more frequently to representations at the sensorimotor level, but fruitful parallels to later stages of development can be made. "Scheme" is used to refer to a type of *representation of actions that is general and does not include the particulars of the action or the objects that it is applied to.* For example, "grasping" as a scheme of action is a generalized action and does not involve any particulars of the action (such as the position of the hands and fingers) or any representation of the grasped objects. "Schema," in contrast, is used by Piaget to refer to a *representation of the object as a consequence of its assimilation by a scheme of action.* An object is schematized – that is, represented by the child – as it is assimilated by a scheme because of the particular adaptations it requires of the subject in carrying out the action. For example, grasping a ball and grasping a stick require different particular adaptations of the same generalized action of grasping. When the child grasps the ball or the stick, the hands must adapt to the shape of the object. These efforts to adapt form in the subject an impression of the object through the scheme of action being used. Thus, the use of a *scheme* results in the development of a *schema* of the object to which the scheme is applied. By having to adapt to new objects, the scheme becomes more general and flexible.

At a more advanced level of development, schemes of actions and their coordination give origin to logicomathematical structures. These structures are representations of logicomathematical relations without commitment to any particular content. In Piagetian theory, these logicomathematical structures are not the content of the subject's thinking but a way of thinking. The structure of classification, for example, involves the application of criteria for classifying in an exhaustive way, generating mutually exclusive classes. As a general structure, the particulars of the criteria used to classify objects do not matter. However, people do not *think about* these principles of classification in their everyday activities; instead, they *use* the principles in generating and learning *classification systems*. The structure of classification contrasts with knowledge of systems of classification, such as a system for classifying flowers, for example, because the criteria used in the classification are irrelevant for the structure while being of central importance for the classification system. Systems of classification contain information about the particular criteria used to classify the objects and about the organization of the system. In a system of classification, it does matter which criteria are applied.

In a general way, systems of knowledge learned in everyday life, like measurement and monetary systems, correspond – at a more advanced level – to the schemas of the sensorimotor period. They include abstract logicomathematical relations and lived-through situations in their representation. Knowledge of a monetary system used in everyday life includes knowledge of logicomathematical principles (of units, additive composition of totals, etc.) and also of the particular coins and notes that are part of the currency. In contrast, school mathematics is meant to look beyond the particulars of situations and make the logicomathematical principles that structure the systems of knowledge into objects of thought.

Street mathematics involves people in understanding mathematical relationships that are embedded in particular activities, technologies, and situations. *In order to function well in these cultural contexts, subjects must understand the mathematical invariants as well as the particulars of the situations.* For example, if subjects from another culture understand all the invariants involved in a system of measurement but ignore the particulars of the system used in the new culture – like how many inches make a foot and how many feet make a yard – the subjects are *in theory* able to understand the measurement system in the new culture. However, they may not be able to use it when they first get there. In order to use any measurement system, one must know both the general logic of measurement and the specific operations of measuring in that system. Knowing a measurement system does not mean becoming aware of the principles of measurement but using these principles in practice. The general principles may be understood insofar as they structure the particular measurement practice but may not be an object of consideration in and of themselves. For example, subjects may know that 100 cm make 1 meter without reflecting that there must be an invariant relationship between different units within a system. Subjects know the principles of measurement as they are instantiated within the system.

Does this analysis lead to the conclusion that street mathematics is "particular," whereas school mathematics is "general"? The studies we discussed in the preceding chapters indicate that there is a need to make the distinction between general and particular forms of knowledge less rigid. Most of the foremen were clearly able to work with unknown scales; they displayed knowledge of scales in general, not simply knowledge of the particular scales they had dealt with. Similarly, fishermen showed their ability to solve proportions problems within the isomorphism of measures type of situation with numbers that contradicted their experience and with variables that did not belong to their practice. The polarization of forms of knowledge into general and particular may need some reformulation.

In the next section, we discuss an approach to the analysis of reasoning in logic problems that attempts exactly that: to classify forms of knowledge in a different way, one that does not end in the polarization between general and particular knowledge.

Pragmatic reasoning schemas: Cheng and Holyoak (1985) proposed the notion of *pragmatic reasoning schemas* in an attempt to understand reasoning about propositions of the *if–then* type. The truth of such propositions can be tested in an algorithmic fashion irrespective of content. "If p then q" is a proposition that can be falsified only by cases where p is true but q is not. This general form sets aside the content of the propositions, because content is seen as irrelevant to the logical form – just as the objects that numbers refer to are irrelevant to the results of arithmetic operations. Thus, it was expected that subjects who were capable of handling, *if–then* propositions with one content would be equally competent when reasoning about other contents expressed in the same form. However, research into logical reasoning did not seem to support this expectation: Adults' performance in reasoning tasks involving *if–then* propositions was shown to vary markedly across contents. In the next paragraphs, we describe the experimental paradigm used in these studies and the solution offered by Cheng and Holyoak to explain through the notion of pragmatic reasoning schemas the apparent contradiction observed in the results.

The experimental paradigm that originated this line of research was developed on the basis of a well-known task designed by Wason (1966) to investigate people's reasoning about *if–then* propositions. Wason attempted to make the content of the propositions as divorced from reality as possible in order to avoid an influence from empirical knowledge on performance. In the Wason task, subjects are shown four cards, each of which has a letter on one face and a number on the other. The cards are placed on the table, and thus subjects can see only one face. They are asked to verify whether the conditional rule "If the card has a vowel on one side, then it has an even number on the other side" was followed when the cards were prepared. The cards were laid on the table in such a way that the visible side of one had a vowel, of another a consonant, of a third an odd number, and of the last an even number. The most economical test of whether the rule was violated would require the subjects to turn only two cards: the card with the vowel and the card with the odd number. Only these two cards could falsify the rule. This solution, however, has proved rather difficult for adults (Wason, 1966). Most subjects select the cards with the vowel and with the even number, and fail to select a card that could falsify the rule, namely the card with the odd number. As a consequence of the high rate of errors observed even among college students, the Wason task became the focus of much research that aimed at understanding the failure rates among educated adults.

One of the investigations subsequently carried out (Johnson-Laird, Legrenzi, & Legrenzi, 1972) showed that success rates in this task were much higher if, instead of checking for the application of a totally arbitrary rule, such as the one above, subjects had to check for the application of a sensible rule that (at that time) applied in their everyday life. The rule to be checked by subjects stated that if a letter is sealed it must have a first-class stamp. Subjects were shown four envelopes on a table, two of which displayed the stamps and their value, and two of which had the stamps facing down so that the subject was able to see whether or not these letters were sealed. Just as was required in the original Wason task, subjects had to turn two letters in order to test for the falsification of the rule: the sealed letter and the one displaying a second-class stamp. Despite the similarity in logical structure and in the way the task was carried out, performance proved to be context-sensitive – that is, subjects did significantly better in this new version of the task than in the original Wason task. Although variations were observed across studies, with some studies yielding significant differences and others not, the effect across studies was consistently in the direction of facilitation of the task when the problem was presented in ways that made "human sense" (Donaldson, 1978) to the subjects.

The contrast between solving propositional problems about arbitrary *if–then* rules, on the one hand, and solving problems in which *if–then* rules are part of sensible situations, on the other hand, parallels the school–street mathematics distinction. In street mathematics, problem-solving activities are carried out in situations that are part of everyday life. As pointed out earlier, successful learning and problem solving in everyday life may be explained by the preservation of meaning during problem-solving activities. Subjects who solved the modified version of the Wason task in which the content of the conditional rule referred to sealed letters and the value of stamps certainly understood the conditions under which violating the rule had a consequence. They knew that overpaying was of no consequence and did not consider these cases in their test of the rule, whereas underpaying resulted in the letter being returned to the sender. They could use a socially meaningful representation of the situation to understand the problem.

Are these situational schemas highly specific, useful only in the closely defined set of circumstances in which they developed? Or is it possible to conceive of these situational schemas as having some generality? Did subjects who performed well with "sensible" rules but poorly with "arbitrary" rules really understand propositional logic, or did they simply, when successful, use falsification rules that applied in real situations?

Cheng, Holyoak, Nisbett, and Oliver (1986) recently summarized the debate in this area by arguing that two views have dominated theories of deductive reasoning. One view holds that people use syntactic, domain-independent rules of logic in reasoning. The second, developed partly in response to the empirical difficulties faced by the first, "holds that people do not use syntactic rules in reasoning but instead develop much narrower rules tied to particular content domains in which people have actual experience. Such specific rules, or perhaps simple memory of examples and counterexamples, are used to evaluate the validity of propositions" (p. 293). In contrast to these two views, Cheng & Holyoak (1985, p. 395), proposed that

people often reason using neither syntactic, context-free rules of inference, nor memory of specific experiences. Rather, they reason using abstract knowledge structures induced from ordinary life experiences, such as "permission," "obligations," and "causation." Such knowledge structures are termed *pragmatic reasoning schemas.* A pragmatic reasoning schema consists of a set of generalized, context-sensitive rules which, unlike purely syntactic rules, are defined in terms of classes of goals (such as taking desirable actions or making predictions about possible future events) and relationships to the goals (such as cause and effect or precondition and allowable action).

Cheng and Holyoak's notion of pragmatic schemas, developed in relation to the *if–then* rules in logic, makes new theoretical contributions that we can relate to the Piagetian concept of scheme and schema. A schema of an object is rather particular; a pragmatic schema applies in a variety of situations that share the same definition. Knowledge structures of social relations (such as permission, prohibition, promise, etc.) and of empirical relations (such as causation) as conceived in their theory are not limited and bound to particular experiences but relate to classes of events with a common social and logical organization. Pragmatic schemas are a general way of understanding new situations and new experiences – just like Piagetian abstract reasoning structures – but they are context-sensitive, not context-free. For example, the pragmatic schema of prohibitions is not the knowledge of one social rule but of a class of social rules that may (or may not) invoke conditions of application. For example, a rule like "Children are not allowed out of the classroom during work time" invokes a condition of application – that is, if you are a child, you are not allowed out of the classroom but if you are a teacher the rule does not apply. Thus, understanding prohibition involves understanding *if–then* statements that are part of the "logical" aspects of the situation. This pragmatic schema can be used in a variety of situations, allowing people to make sense of the social and logical factors that are expected to regulate behavior in those situations.

Girotto, Gilly, Blaye, and Light (1991) were able to show that when 10–14-year-old children interpreted a statement as a prohibition[1]

---

[1] Girotto et al. discuss their example as a permission rather than a prohibition rule, but this difference is not relevant for present purposes.

rule, they were much more successful in testing whether the rule had been violated than if the rule could not be interpreted as a prohibition (being simply an arbitrary rule). Success rates in the prohibition interpretation of a rule were as high as 80%, regardless of whether the rule was familiar ("Passengers in the front seat of a car must wear seat belts") or unfamiliar ("Cars must be painted in fluorescent colors if they are going to be driven at above 100 km per hour"). This contrasts strongly with the 10% correct responses observed when the traditional Wason task was given – a task involving a rule that does not allow for the use of pragmatic schemas from everyday life.

The idea of pragmatic schemas developed in the context of logical reasoning can easily be applied to the analysis of everyday mathematics. Many everyday activities clearly embed logicomathematical concepts that can be used in a general way even though they may not involve context-free reasoning. The use of a monetary system is one example. Counting and calculating with money involve representing money totals as constant even if coins of different values are used in composing the total. This knowledge can be termed "additive composition of money," paraphrasing the additive composition of number. Additive composition could be a general logico-mathematical principle understood in the context of dealing with money.

Clearly, the way in which additive composition principles are used differ when people are calculating with money and calculating with paper and pencil. For example, composing a total of 80 cents may be a matter of putting together three quarters and one nickel. In contrast, writing 80 cents involves composing a number with eight tens and zero units. Both representations respect the principle of additive composition, but the representation of the money total may *evoke* this principle much more clearly than does a written number. Perhaps the power to evoke the sense of the situation is one of the reasons for the discrepancies in performance we observed between street markets and schools. It is possible that successful subjects not only have a schema of the situation in which the problem is being solved but can also keep it in mind more clearly as a consequence of the way the situation is represented. Thus, we may need not only theoretical ideas that overcome the polarization between general and particular knowledge but also ideas that bring to the fore the importance of forms of representation in thinking. Just such a theoretical approach is discussed in the next section.

Vergnaud's theory about concepts: A third theoretical approach we want to consider here is Vergnaud's (1985) theory about the analysis of mathematical concepts. A mathematical concept (or model), according to Vergnaud (1985), always involves a *set of situations* that give meaning to the concept, a *set of invariants* that are constituted by the relationships essential to the concept, and a *set of symbols* that are used in the representation of the concept. Vergnaud's theory thus considers both the structure and the content of a concept, like the preceding view, but considers a further element: symbolic representation. Vergnaud points out that representations always involve keeping some features of the concept in focus while losing sight of others. Thus, the way in which symbols are used in the representation of the concept determines what is represented and what is not represented at particular moments. Theories about the development of mathematical concepts have turned mainly to an understanding of invariants in order to characterize differences in concepts. In the discussion of street and school arithmetic, however, we saw that looking at invariants is not enough to characterize the differences between the two practices. Children draw on the same invariants in and out of school but rely on different types of representation. In the written forms of representation typically used in school, the meaning of the situation is purposely not represented. It is by not representing the referents that it becomes possible, for example, to multiply across measures – that is, multiply money by number of coconuts and find a result in money. It is also by ignoring the relative value of digits that it is possible to proceed with column arithmetic, as we do in the written algorithms. As the situational aspects are lost in written representation, syntactic rules become the focus. Conversely, oral arithmetic maintains in focus the meaning of the situations, and less attention is given to syntactic rules. This difference in representation has clear consequences for problem solving. As we saw earlier, both rates of correct response and types of error are different in oral and written arithmetic.

Keeping the meaning of situations in perspective seems to help not only with carrying out arithmetic but also with using situational schemas for solving more complex problems. Through the analysis of complex problems discussed in chapter 4, it became clear that farmers and carpenters were more able than students to use their computation skills to solve problems involving several steps. To use Cheng and Holyoak's terminology, their pragmatic schemas allowed them to organize relationships in the experimental tasks much better than students could with knowledge of mathematics built in the classroom. Farmers, for example, calculated the number of tea bushes needed to plant an area fully by repeatedly using an isomorphism of measures conception. They generated three-step scalar solutions, calculating first how many bushes in a row, then how many rows in the area, and finally how many bushes altogether. The performance of fifth-grade students, in contrast, broke down when problems involving more than one step in calculation were presented. Although they knew the necessary arithmetic – multiplication and division – and had solved two-variable proportions problems using scalar solutions, they did not accomplish a meaningful analysis of this more complex situation. Similarly, apprentices in a carpentry shop showed a breakdown in performance when they had to analyze relationships between quantities in more complex problems, whereas professional carpenters, who used oral arithmetic practices, were able to preserve the meaning of the situation as they carried out their calculations.

The role of representations in the preservation of situational meaning and the consequent use of pragmatic mathematical schemas in problem solving were also explored by Nunes (1993) in a study about directed numbers. In this study, students were randomly assigned to either an oral or a written condition of testing for solving problems with directed numbers. The problems were about a farmer's profits and losses on his different harvests. Students at three grade levels (fourth, fifth, and sixth) did significantly better in the oral than in the written condition, despite the fact that they had been randomly assigned to the conditions so that the use of oral or written arithmetic did not reflect their own choice. Moreover, some students who arrived at the wrong answer in the written condition were able to provide the correct answer immediately afterward when asked to explain their procedure. In the process of providing oral explanations, they did use their pragmatic schemas about debts and profits and rejected their previous written response, which had followed rules for calculating with directed numbers learned only in part. Despite their ability to understand the invariants in the situation and to solve the problem orally, their initial recourse to written representation and the disconnection between their street mathematics and their school mathematics led them astray.

In summary, street mathematics involves people in understanding particular activities, technologies, and situations. People form representations of these particular situations that clearly contain elements of the situation but that may nevertheless be treated in a more general fashion – as a model for several situations – even at the level of elementary addition. By solving everyday problems about addition, people develop pragmatic schemas of addition – neither a general theory of addition nor simply knowledge of addition bonds. The fact that specific information is contained in representations in street mathematics is not a drawback. It is specific information that allows subjects to control for the meaning and reasonableness of their answers in problem situations. It was clearly demonstrated in the studies about proportions that foremen and fishermen can use their street mathematics knowledge in new situations that involve the same logicomathematical relationships. It was also observed that level of schooling had no effect on whether the knowledge developed outside school could be used in new ways. Thus, representation of the particulars of the situation does not imply that the subject is restricted to understanding that situation. There is ample evidence for flexibility and generalizability of the pragmatic schemas of street mathematics. There is also plenty of evidence of the disconnection between people's knowledge of street and of school mathematics. Such findings point to the need to consider some lessons for education.

II Lessons for education

The lessons for education we shall be considering here bear on two points. First, we consider the idea of an approach to mathematics teaching that seeks its inspiration in street mathematics, and then we discuss the interaction between street and school mathematics in the development of critical thinking.

1. Street mathematics in the classroom

One of the lessons about street mathematics that we can take to the classroom is about preserving meaning during the mathematizing of situations. This idea has been extensively explored under the concept *realistic mathematics education,* most developed among researchers of the Freudenthal Institute in Utrecht. Realistic mathematics education involves posing to pupils in the classroom problems that require the consideration of empirical constraints as well as social and logical rules that apply outside school. In this conception, solving a problem involves making decisions about how to proceed in imagined situations. Mere exercises for the application of procedures learned in the classroom are not part of this notion. The contrast between the two problems in Figure 7.1 is instructive. It might be said that they are both simply a matter of using one operation. There are, however, important differences between them. The first problem was used to introduce division and presents a situation that can be imagined. Its solution can lead to a decision about how many tables are needed in the school hall tonight. Mathematics is a tool for understanding a situation. The need to make a decision, albeit in an imagined situation, gives sense to the problem. It is a situation in which one would use mathematics outside the classroom. The second problem was used as a task in an experiment to test children's activation of their knowledge of the world in solving word problems. However, the situation described clearly goes against the idea of what it is sensible to do when going to a fast-food place. Its solution is not meant to be used in reaching any decision. Solving the multiplication question does not help anyone understand a situation that may arise in everyday life. One can hardly imagine such a situation emerging in everyday life and our using mathematics in it.

The contrast between the two problems illustrates the difficulty of thinking of good problems to work from in the classroom. However, if mathematics education is going to be realistic, problems will have to be sought that respect assumptions about life outside school. As Gravemeijer (1990) pointed out, some problems that have easy mathematical solutions are in fact jokes, because the constraints of life are ignored in the problem-solving process. He illustrates this point with this problem: There were 10 birds in a tree; 2 were killed by a gunshot; how many were left? The arithmetic answer "8" violates assumptions about the outside world – all living birds would have flown away.

The role of problem situations in realistic mathematics education is central. From problem situations, students are expected to work on understanding relationships between variables, to develop models,

1. A problem from realistic mathematics education
Tonight 81 parents will visit our school.
At each table 6 parents can be seated.

How many tables do we need?

2. An unrealistic problem about time
John ate 8 Big Macs. It takes John 15 minutes to eat a Big Mac.
How long did it take him to eat them all?

7.1 A single operation, two arithmetic problems. *Sources:* (1) Gravemeijer, 1990; (2) Baranes, Perry, & Stigler, 1989

and to be able to proceed from their "own informal mathematical constructions to what could be accepted as formal mathematics" (Streefland, 1990, p. 1).

"The realistic approach rejects the instructional partition between mathematical knowledge and its application. On the contrary, applications play an important role in the factual development of mathematical concepts and skills" (Gravemeijer, 1990, p. 10). According to the proposers of realistic mathematics education, models built in the classroom about problem situations arise from problem-solving activities. For this reason, the models can work as a bridge between street and school mathematics. Students can start from their own representations of the situation and be provided with mathematical tools that help them connect several aspects of their knowledge. For example, Gravemeijer observed that when children solved the realistic problem presented in Figure 7.1 they made connections between counting, addition, multiplication, and division. Some children drew small rectangles to represent tables, wrote the number "6" inside each rectangle, and then counted in 1s or 6s, up to 81. At the end of this process, it was necessary only to count the tables. A second solution involved figuring out that 10 tables would seat 60 persons and then using combinations of addition and counting to figure out how many more tables were needed to seat 81 persons. A third type of solution was displayed by one student who knew that $6 \times 6 = 36$, doubled this value to obtain 12 tables seating 72 people, and then

Table 7.1

---

81 persons

$10 \times 6 = 60$ 81
                      -60
                      ——
                      21

21 persons left to find seats
6, 12, 18 – 3 tables will be full, and one more table will have 3 persons

Note that the procedure used here is to approach the dividend by finding factors of the divisor that do not exceed it. The treatment given to the remainder is consistent with the question posed.

---

counted the number of tables needed to get the remaining people seated. These students had the chance to use their own resources to make sense of the problem – to start from their own mathematics. By comparing their methods, they could see that one method was more efficient than another. They learned both the equivalence of different solutions and something about seeking efficiency. It is amazing how close the second solution actually comes to putting multiplication and division together and to rediscovering long division – yet many teachers may not notice that fact without the support of a more formal representation, such as the one presented in Table 7.1. The problem here is that teachers themselves have grown up with a gap between informal and formal mathematics.

Realistic mathematics education does not only start from problems and allow pupils to use their own representations in problem solving. It also relies on a (partially implicit) theory of situations and the pragmatic schemas built in the process of understanding them, because one of the explicit aims of this approach is to seek coordinations between different schemas and forms of representation. For example, children may be asked to solve a division problem presented pictorially that involves distributing in a fair way to 3 children 36 pieces of candy displayed in a square array. The children reproduce the situation on their own paper by drawing the children and their corresponding pieces. As the pieces are distributed, they are canceled from the array. Some children aim at forming three equal groups, using a sharing schema. They give each child 1 piece in turn. Other children draw groups of three – in each round, three pieces are given away – using a schema that controls for the size of the stock of pieces and the number needed in each round. Instruction aims at the coordination of these two main solutions, termed "distribution" and "ratio division" (Gravemeijer, 1990), respectively. This example shows the importance of a theory of situations and the analysis of pragmatic schemas in realistic mathematics education. Although much progress in this respect has already been accomplished through a decade of research (see, for example, Carpenter & Moser, 1982; Hart, 1981; Riley, Greeno, & Heller, 1983; and Vergnaud, 1982, 1983), there are still many open questions.

2. Street mathematics, school mathematics, and critical thinking.

The pragmatic schemas of street mathematics in some sense take reality for granted. A linear model is *assumed* when one shops in a street market. A similar relationship is also assumed for the relationship between the price of production and price to the consumer. Often when price comparisons are carried out across brands of a particular product, deviations from the linear model are "explained by" the better quality of the more expensive product. If one questions a salesperson about the difference in price between, for example, two washing machines that offer basically the same options but carry different brand names, one often hears that the more expensive one must be of better quality. Not disputing the fact that this may be (but not always is) the case, it may still not cost more to produce the better machine. The explanation may be in a manufacturer's expectation of larger profits, in a desire to maintain the prestige of higher-quality products, or in less efficient production by one of the manufacturers. In the same way that we assume that certain models fit certain situations, we also assume that there are justifications – empirical, logical, and social – for the model to be used. This was seen to be the case in the study by Girotto and his associates (1991) in which an *if–then* rule arbitrarily invented by the experimenters for the task was interpreted as meaningful regardless of how the conditions were set up – speed limits of under 100 km per hour were seen as justified by children irrespective of the application to fluorescent or nonfluorescent cars. Despite a statistically significant difference in favor of speed limits for nonfluorescent cars, half the children in the experiment found ways of justifying speed limits set in precisely the opposite direction. In other words, they could find social justifications for a rule regardless of its direction. They assumed that the social rules made sense, taking social reality for granted – perhaps because the rules were introduced in an experimental situation where the person in charge was assumed to be reasonable.

Many decisions with grave social consequences are based upon *assumed* mathematical models. A critical mathematics education would involve making these models explicit and questioning them by raising the possibility of other socially defined realities. What is a fair way to set income tax bands? How are salaries calculated? What adjustments should be made to the wages of workers when the prices of the objects they produce rise? What determines the rate of interest paid by home buyers? How is the level of inflation calculated? Why is the value of wages in one country one tenth of the value of wages in another country, when workers are producing the same products?

There is a sense in which the answers to all these questions are obtained by mathematical models. *Once the mathematical model is assumed,* the answer is straightforward. However, the models must be at the core of social negotiations and do not have to be taken for granted. An exercise that is central to the development of critical thinking involves the differentiation of the social, logical, and empirical constraints of a situation. Mathematics education can play a key role in the development of critical thinking by bringing into focus the effects of adopting different mathematical models to define the same social situation. Testing the impact of different models on the lives of people may affect the way in which mathematical models will be viewed as part of the social and logical rules of situations. A useful exercise that we have carried out with mathematics teachers relates to the computation of inflation indexes. At several steps in this computation, choices are made about the mathematical model to be used. For example, which items should be taken into account? How will they be weighted in the final equation? It was a useful exercise to allow teachers to put in and take out factors, weight different factors separately, and look at different equations used by the government of the same country at different times; they came to the realization that different ways of mathematizing the same situation have drastically different results. There was no question of mathematical error in these cases. Rather, the question centered on the social sense of the mathematics used. Can there be a more appropriate place for such discussion than a mathematics class? Will the relationship between mathematical models and situations not be considerably clarified when there is no empirically correct answer to be discovered, but only sound, sensible answers to be constructed?

III Final comments

We started this study with the idea that street mathematics was a phenomenon of social significance. We recognized that street mathematics plays a major role in societies that do not give official recognition to the mathematical abilities actually needed by people every day both in and outside their jobs. We searched for subjects who had the opportunity to develop knowledge of street mathematics beyond their knowledge of school mathematics. Street mathematics has often been treated in the literature as "lesser mathematics" involving idiosyncratic, intuitive, childlike procedures – techniques that did not allow for generalization and should thus be eliminated in the classroom through carefully designed instruction. We were able to document the fact that street mathematics is not the learning of particular procedures repeated in an automatic, unthinking way, but involves the development of mathematical concepts and processes. Street mathematics represents one example of cognition in practice, as Lave (1988) calls it. It is a type of thinking that is carried out under the constraints of social, empirical, and logical rules and that aims at accomplishing a result. Study of street mathematics revealed the existence of considerable potential for learning and understanding mathematical concepts in people who have often been treated as unfit to learn mathematics in school.

We have not been alone in recognizing the value of street mathematics. Realistic mathematics educators have been exploring ways of teaching that treat formalization in mathematics as a process of obtaining progressively more formal representations of mathematical understanding built outside school. Children are asked to solve in the classroom imagined problems to which they must apply the same social and empirical constraints that apply outside the classroom. They are asked to interpret the results of their problem-solving efforts in this same fashion, considering empirical and social constraints, not in formal terms. By proceeding in this way, realistic mathematics education appears to be successful in bringing children to build their knowledge of school mathematics on the foundation of their already available knowledge of street mathematics.

Uncovering people's considerable potential for understanding mathematics outside the classroom raises several questions for researchers in both psychology and education. As we looked at the results of the different studies in the chapters above, we realized the inadequacy of the psychological theories that have been the basis of pedagogies in mathematics. Most theories have only looked at the logical constraints for solving problems. Comparatively little effort has been invested in analyzing the types of models that people build to understand naturally occurring or meaningful imagined situations and the varied forms of representation that emerge in these model-building efforts. The challenge for researchers is to look at these neglected aspects of cognition and try to build them into their theories of learning. The challenge for educators is to turn this potential for learning into conscious routes for teaching.

References

Baranes, R., Perry, M., & Stigler, J. W. (1989). Activation of real-world knowledge in the solution of word problems. *Cognition and Instruction, 6,* 247–318.

Berlinck, M. T. (1977). *Marginalidade social e relações de classe em São Paulo.* Petrópolis: Vozes.

Bishop, A. J. (1983). Space and geometry. In R. Lesh & M. Landau (Eds.), *Acquisition of mathematics concepts and processes.* (pp. 175–206). New York: Academic Press.

Blij, F. van der. (1985). Mathematics and the visual arts. In L. Streefland (Ed.) *Proceedings of the Ninth International Conference for the Psychology of Mathematics Education: Vol. 2. Plenary addresses and invited papers* (pp. 15–31). Utrecht: University of Utrecht, Research Group on Mathematics Education and Educational Computer Center.

Brown, J. S., & Burton, R. R. (1978). Diagnostic models for procedural bugs in basic mathematical skills. *Cognitive Science, 2,* 155–192.

Bruner, J. S. (1966). *Toward a theory of instruction.* Cambridge, MA: Harvard University Press.

Bryant, P. E. (1985). The distinction between knowing when to do a sum and knowing how to do it. *Educational Psychology, 5,* 207–215.

Carpenter, T. P., Lindquist, M. M., Matthews, W., & Silver, E. A. (1983). Results of the third NAEP mathematics assessment: Secondary school. *Mathematics Teacher, 76,* 652–659.

Carpenter, T. P., & Moser, J. M. (1982). The development of addition and subtraction skills. In T. P. Carpenter, J. M. Moser, & T. A. Romberg (Eds.), *Addition and subtraction: A cognitive perspective* (pp. 9–24). Hillsdale, NJ: Erlbaum.

Carraher, D. W. (1985a). *Experiments on measurement and estimates in the classroom.* Mestrado em Psicologia da Universidade Federal de Pernambuco, Recife. Unpublished research report.

Carraher, T. N. (1985b). The decimal system: Understanding and notation. In L. Streefland (Ed.), *Proceedings of the Ninth International Conference for the Psychology of Mathematics Education* (Vol. 1, pp. 288–303). Utrecht: University of Utrecht, Research Group on Mathematics Education and Educational Computer Center.

155

———. (1986a). From drawings to buildings: Working with mathematical scales. *International Journal of Behavioural Development, 9,* 527–544.

———. (1986b). *Intuitive knowledge about rational numbers.* Paper presented at the 38th Commission Internationale pour l'Etude et l'Enseignement des Mathématiques, Southampton, England, July.

Carraher, T. N., Carraher, D. W., and Schliemann, A. D. (1985). Mathematics in the streets and in schools. *British Journal of Developmental Psychology, 3,* 21–29.

———. (1986). Proporcionalidade na educação científica e matemática: Desenvolvimento cognitivo e aprendizagem. *Revista Brasileira de Estudos Pedagógicos, 67,* 586–602.

Carraher, T.N., & Schiemann, A.D. (1983). Fracasso escolar: Uma questão social. *Cadernos de Pesquisa,* São Paulo (45): 3–19.

———. (1985). Computation routines prescribed by schools: Help or hindrance? *Journal for Research in Mathematics Education, 16,* 37–44.

Cavalcanti, C. (1978). *Viabilidade do setor informal: A demanda de pequenos serviços no Grande Recife.* Recife: Instituto Joaquim Nabuco de Pesquisas Sociais.

Cavalcanti, C, & Duarte, R. (1980a). *A procura de espaço na economia urbana: O setor informal de Fortaleza.* Recife: SUDENE/FUNDAJ.

———. (1980b). *O setor informal de Salvador: Dimensões, natureza, significação.* Recife: SUDEN/FUNDAJ.

Cheng, P. W., & Holyoak, K. J. (1985). Pragmatic reasoning schemas. *Cognitive Psychology, 17,* 391–416.

Cheng, P. W., Holyoak, K. J., Nisbett, R. E., & Oliver, L. M. (1986). Pragmatic versus syntactic approaches to training deductive reasoning. *Cognitive Psychology, 18,* 293–328.

Cockcroft, W. H. (1986). Inquiry into school teaching of mathematics in England and Wales. In M. Carss (Ed.), *Proceedings of the Fifth International Congress on Mathematics Education,* (pp. 328–329). Boston: Birkhauser.

Cole, M., Gay, J., Glick, J., & Sharp, D. (1971). *The cultural context of learning and thinking.* New York: Basic.

Cole, M., & Scribner, S. (1974). *Culture and thought: A psychological introduction.* New York: Wiley.

Costa, M. (1984). A educação e suas potencialidades. In H. Levin, M. Costa, C. Solari, M. Leal, G. Miranda, & J. Velloso (Eds.)., *Educação e desigualdade no Brasil* (pp. 41–70). Petrópolis: Vozes.

D'Ambrosio, U. (1984). *Ethnomathematics.* Paper presented at Fifth International Congress on Mathematics Education, Adelaide, August.

———. (1986). *Da realidade à ação: Reflexões sobre educação e matemática.* Campinas: Summus Editorial.

Dias, M. G. (1988). *Logical reasoning.* Unpublished doctoral dissertation, Oxford University, Oxford.

Dias, M. G., & Harris, P. L. (1988). The effect of make-believe play on deductive reasoning. *British Journal of Developmental Psychology, 6,* 207–221.

Doise, W., & Mugny, G. (1981). *Le développement social de l'intelligence.* Paris: Interéditions.

Donaldson, M. (1978). *Children's minds.* London: Fontana.

Fahrmeier, E. (1984). Taking inventory: Counting as problem solving. *Quarterly Newsletter of the Laboratory of Comparative Human Cognition, 6,* 6–10.

Fuson, K. C. (1982). An analysis of the counting-on solution procedure in addition. In T.P. Carpenter, J. M. Moser, & T. A. Romberg (Eds.), *Addition and substraction: A cognitive perspective* (pp. 67–82). Hillsdale, NJ: Erlbaum.

Gal'perin, P. Ya., & Georgiev, L. S. (1969). The formation of elementary mathematical notions. In J. Kilpatrick & I. Wirszup (Eds.), *Soviet studies in the psychology of learning and teaching mathematics: Vol. 1. The learning of mathematical concepts* (pp. 189–216). Chicago: University of Chicago Press.

Gay, J., & Cole, M. (1967). *The new mathematics and an old culture: A study of learning among the Kpelle of Liberia.* New York: Holt, Rinehart & Winston.

Ginsburg, H. P. (1977). *Children's arithmetic: The learning process.* New York: Van Nostrand.

——— . (1982). The development of addition in contexts of culture, social class, and race. In T. P. Carpenter, J. M. Moser, & T. A. Romberg (Eds.), *Addition and subtraction: A cognitive perspective* (pp. 99–116). Hillsdale, NJ: Erlbaum.

Girotto, V., Gilly, M., Blaye, A., & Light, P. (1991). *Children's performance in the selection task: Plausibility and familiarity.* Unpublished research report, Open University, U.K.

Grando, N. I. (1988). *A matemática na agricultura e na escola* [Mathematics in agriculture and in school]. Unpublished master's thesis, Universidade Federal de Pernambuco, Recife.

Gravemeijer, K. (1990). Context problems and realistic mathematics education. In K. Gravemeijer, M. van den Heuvel, & L. Streefland (Eds.), *Context free productions tests and geometry in realistic mathematics education* (pp. 10–32). Culemborg Technipress.

Greenfield, P. M. (1966). On culture and conservation. In J. S. Bruner, R. R. Oliver, & P. M. Greenfield (Eds.), *Studies in cognitive development* (pp. 225–256). New York: Wiley.

Groen, G., & Resnick, L. B. (1977). Can preschool children invent addition algorithms? *Journal of Educational Psychology, 79,* 645–652.

Hart, K. (1981). *Children's understanding of mathematics: 11–16.* London: Murray.

——— . (1984). *Ratio: Children's strategies and errors.* Windsor, England: NFER–Nelson.

Hatano, G. (1982). Cognitive consequences of practice in culture-specific procedural skills. *Quarterly Newsletter of the Laboratory of Comparative Human Cognition, 4,* 15–18.

Hughes, M. (1986). *Children and number.* Oxford: Blackwell Publisher.

Hunter, I. M. L. (1977). Mental calculation: Two additional comments. In P. N. Johnson-Lair & P. C. Wason (Eds.), *Thinking: Readings in cognitive science* (pp. 35–42). Cambridge: Cambridge University Press.

Inhelder, B., & Piaget, J. (1958). *The Growth of logical thinking from children to adolescence.* New York: Basic.

Johnson-Laird, P. N., Legrenzi, P., & Legrenzi, M. (1972). Reasoning and a sense of reality. *British Journal of Psychology, 63,* 395–400.

Karplus, R., Karplus, E., Formisano, M., & Paulsen, A. C. (1977). Proportional reasoning and control of variables in seven countries. In J. Lochhead &

158   References

J. Clement (Eds.), *Cognitive process instruction: Research on teaching thinking skills* (pp. 47–104). Philadelphia: Franklin Institute Press.

Karplus, R., Pulos, S., & Stage, E. K. (1983) Proportional reasoning of early adolescents. In R. Lesh & M. Landau (Eds.), *Acquisition of mathematics concepts and processes* (pp. 45–90). London: Academic Press.

Kaye, K. (1982). *The mental and social life of babies.* London: Methuen.

Knight, P. T., & Moran, R. (1981, December). *Brazil.* Poverty and Basic Need Series. Washington, DC: World Bank.

Lakatos, I. (1981). *Matemáticas, ciencia y epistemología.* Madrid: Alianza Editorial.

Lave, J. (1977). Cognitive consequences of traditional apprenticeship training in West Africa. *Anthropology and Educational Quarterly, 7,* 177–180.

————. (1988). *Cognition in Practice: Mind, mathematics and culture in everyday life.* New York: Cambridge University Press.

Lave, J., Murtaugh, M., & de la Rocha, O. (1984). The dialectical construction of arithmetic practice. In. B. Rogoff & J. Lave (Eds.), *Everyday cognition: Its development in social context* (pp. 67–97). Cambridge, MA: Harvard University Press.

Levin, H. M. (1984). Educação e desigualdade no Brasil: Uma visão geral. In H. Levin, M. Costa, C. Solari, M. Leal, G. Miranda, & J. Velloso (Eds.), *Educação e desigualdade no Brasil* (pp. 15–40). Petrópolis: Vozes.

Levy, S. (1969). *An economic analysis of investment in education in the state of São Paulo.* São Paulo: Instituto de Pesquisas Econômicas da Universidade de São Paulo.

Light, P. (1986). Context, conservation, and conversation. In M. Richards & P. Light (Eds.), *Children of social worlds* (pp. 170–190). Cambridge: Polity.

Lunzer, E. A., Harrison, C., & Davey, M. (1972). The four-card problem and the development of formal reasoning. *Quarterly Journal of Experimental Psychology, 24,* 326–339.

Luria, A. R. (1976) *Cognitive development: Its cultural and social foundations.* Cambridge, MA: Harvard University Press.

————. (1979). *Curso de psicologia geral* (Vol. 1). Rio de Janeiro: Civilização Brasileira.

Noelting, G. (1980). The development of proportional reasoning and the ratio concept. *Educational Studies in Mathematics, 1,* 331–363.

Nunes, T. (1993). Learning mathematics: Perspectives from everyday life. In R. Davies & C. A. Maher (Eds.), *Schools, mathematics, and the world of reality* (pp. 61–78). Boston: Allyn & Bacon.

Nunes, T., & Bryant, P. E. (1992). *Children doing mathematics.* Manuscript in preparation, Institute of Education, University of London.

Pastore, J. (1982). *Inequality and social mobility in Brazil.* Madison: University of Wisconsin Press. (Original work published 1979.)

Perret-Clermont, A. N. (1980). *Social interaction and cognitive development in children.* London: Academic Press.

Piaget, J. (1926). *The language and thought of the child.* New York: Harcourt Brace.

————. (1952). *The child's conception of number.* London: Routledge & Kegan Paul.

Piaget, J., Grize, J. B., Szeminska, A., & Bang, V. (1968). *Epistémologie et psychologie de la fonction.* Paris: PUF.

Plunkett, S. (1979). Decomposition and all that rot. *Mathematics in Schools, 8,* 2–7.

Reed, H. J., & Lave, J. (1981). Arithmetic as a tool for investigating relations between culture and cognition. In R. W. Casson (Ed.), *Language, culture and cognition: Anthropological perspectives* (pp. 437–455). New York: Macmillan.

Resnick, L. B. (1982). Syntax and semantics in learning to subtract. In T. P. Carpenter, J. M. Moser, & T. A. Romberg (Eds.), *Addition and subtraction: A cognitive perspective* (pp. 135–155). Hillsdale, NJ: Erlbaum.

———. (1986). The development of mathematical intuition. In M. Perlmutter (Ed.), *Minnesota symposium on child development* (Vol. 19, pp. 159–193). Hillsdale, NJ: Erlbaum.

Riley, M. S., Greeno, J. G., & Heller, J. I. (1983). Development of children's problem solving skills in arithmetic. In H. P. Ginsburg (Ed.), *The development of mathematical thinking* (pp. 306–313). London: Academic Press.

Rogoff, B. (1981). Schooling and the development of cognitive skills. In H. C. Triandis & A. Heron (Eds.), *Handbook of cross-cultural psychology* (Vol. 4). Boston: Allyn & Bacon.

Saxe, G. B. (1982). Culture and the development of numerical cognition: Studies among the Oksapmin. In C. J. Brainerd (Ed.), *Children's logical and mathematical cognition* (pp. 157–176). New York: Springer-Verlag.

Saxe, G. B., & Moylan, T. (1982). The development of measurement operations among the Oksapmin of Papua New Guinea. *Child Development, 53,* 1242–1248.

Saxe, G. B., & Posner, J. K. (1983). The development of numerical cognition: Cross-cultural perspectives. In H. P. Ginsburg (Ed.), *The development of mathematical thinking* (pp. 291–317). New York: Academic Press.

Schliemann, A. D. (1984). Mathematics among carpenters and apprentices. In P. Damerow, M. W. Dunckley, B. F. Nebres, & B. Werry (Eds.), *Mathematics for all* (pp. 92–95). Paris: UNESCO.

Schliemann, A.D., & Nunes, T. (1990). A situated schema of proportionality. *British Jounal of Developmental Psychology, 8,* 259–268.

Schoenfeld, A. H. (1988). When good teaching leads to bad results: The disasters of well taught mathematics courses. *Educational Psychologist, 23*(2), 145–166.

Scribner, S. (1975). Models of thinking and ways of speaking: Culture and logic reconsidered. In P. N. Johnson-Laird & P. C. Wason (Eds.), *Thinking* (pp. 483–500). New York: Cambridge University Press.

———. (1984a). Product assembly: Optimizing strategies and their acquisition. *Quarterly Newsletter of the Laboratory of Comparative Human Cognition, 6,* 11–19.

———. (1984b). Organizing knowledge at work. *Quarterly Newsletter of the Laboratory of Comparative Human Cognition, 6,* 26–32.

———. (1984c). Studying working intelligence. In B. Rogoff & J. Lave (Eds.), *Everyday cognition: Its development in social context* (pp. 9–40). Cambridge, MA: Harvard University Press.

———. (1984d). Pricing delivery tickets: "School arithmetic" in a practical setting. *Quarterly Newsletter of the Laboratory of Comparative Human Cognition, 6,* 19–25.

———. (1986). Thinking in action: Some characteristics of practical thought. In R. J. Sternberg & R. K. Wagner (Eds.), *Practical Intelligence: Nature and origins of competence in the everyday world* (pp. 13–60). Cambridge, MA: Harvard University Press.

Scribner, S., & Cole, M. (1973). Cognitive consequences of formal and informal education. *Science, 182*, 553–559.

————. (1981). *The psychology of literacy.* Cambridge, MA: Harvard University Press.

Sharp, D. W., Cole, M., & Lave, C. (1979). Education and cognitive development: The evidence from experimental research. *Monographs of the Society for Research in Child Development, 44* (1–2, Serial No. 178).

Steffe, L. P., Thompson, P. W., & Richards, J. (1982). Children's counting in arithmetical problem solving. In T. P. Carpenter, J. M. Moser, & T. A. Romberg (Eds.), *Addition and subtraction: A cognitive perspective* (pp. 99–116). Hillsdale, NJ: Erlbaum.

Streefland, L. (1990). Realistic mathematics education: What does it mean? In K. Gravemeijer, M. van den Heuvel, & L. Streefland (Eds.), *Context free productions tests and geometry in realistic mathematics education* (pp. 1–9). Culemborg: Technipress.

Velloso, J. R. (1975). *Human capital and market segmentation: An analysis of the distributions of earnings in Brazil, 1970.* Unpublished doctoral dissertation, Stanford, CA.

————. (1984). Distribuição de renda: Educação e política de estado. In H. Levin, M. Costa, C. Solari, M. Leal, G. Miranda, & J. Velloso (Eds.), *Educação e desigualdade no Brasil* (pp. 173–204). Petrópolis: Vozes.

Vergnaud, G. (1979). Didactics and acquisition of multiplicative structures in secondary school. In W. F. Archenhold, R. H. Driver, A. Orton, & C. Wood-Robinson (Eds.), *Cognitive development research in science and mathematics* (pp. 190–201). Leeds: University of Leeds Press.

————. (1982). A classification of cognitive tasks and operations of thought involved in addition and subtraction problems. In T. P. Carpenter, J. M. Moser, & T. A. Romberg (Eds.), *Addition and subtraction: A cognitive perspective* (pp. 39–59). Hillsdale, NJ: Erlbaum.

————. (1983). Multiplicative structures. In R. Lesh & M. Landau (Eds.), *Acquisition of mathematics concepts and processes* (pp. 128–175). London: Academic Press.

————. (1985). Concepts et schèmes dans une théorie opératoire de la représentation. *Psychologie Française, 30*(3–4), 245–252.

————. (1987). About constructivism. In J. C. Bergeron, N. Herscovics, & C. Kieran (Eds.), *Proceedings of the Eleventh International Conference, Psychology of Mathematics Education* (pp. 42–54). Montreal: University of Montreal Press.

Vygotsky, L. S. (1962). *Thought and language.* Cambridge, MA: MIT Press.

Wason, P. C. (1966). Reasoning: In B. Foss (Ed.), *New horizons in psychology* (pp. 135–151). Harmondsworth: Penguin.

Wason, P. C., & Shapiro, D. (1971). Natural and contrived experience in a reasoning problem. *Quarterly Journal of Experimental Psychology, 23*, 63–71.

Young, R. M., & O'Shea, T. (1981). Errors in children's subtraction. *Cognitive Science, 5*, 153–177.

Author index

Subject index